LIANE
A CAT FROM THE WILD

For Ann Coulson, a
faithful reader.

Mike Tomkies

FOR MARGOT

Other books by the author

The Big Man

It Sure Beats Working

Alone in the Wilderness

My Wilderness Wildcats

LIANE

A CAT FROM THE WILD

MIKE TOMKIES

MACDONALD & JANES · LONDON

First published in 1979 by
Macdonald and Jane's Publishers Ltd
Paulton House
8 Shepherdess Walk
London N1 7LW

ISBN 0 354 04374 9

Filmset in 'Monophoto' Times New Roman by
Servis Filmsetting Ltd, Manchester

Printed in Great Britain by
Purnell & Sons Ltd, Paulton (Bristol) and London

CHAPTER 1

Sylvesturr would greet me with a devastating spit and foot stamp when I took him his food. Then he would often lick his lips in anticipation of the meal ahead.

Outside the night has fallen in a dead calm, bathing the world around us in deepest darkest blue, and the only sound through the partially open window is the faint whispering of the waterfall in the woods. I am sitting in my old keeper's cottage at Wildernesse below the mountains on the remote shores of a Scottish Highland loch, the paraffin lamp flooding the log fireplace and the white, book-lined walls with soft golden light. On the bed to my right a wildcat sleeps, her face sweet and gentle in repose, her white furred mouth and chin peeping between her long forelegs which are stretched out in front of her in an attitude of luxurious ease. I give a slight '*Mau*' and instantly she wakes, her big golden eyes look into mine and she responds as usual with the little chirring trill, deep in her throat, which I have come to know as a greeting of affection. On the floor below her the huge head of my only other companion, a large Alsatian called Moobli, looks up briefly from his sleep, the yellow tasselled bed cover draped comically over his eyes. Now the wildcat stretches her forelimbs out even further, the gleaming goldy-velvet toes spread wide apart, the thick claws digging into but not harming a single thread of my counterpane. They retract again and as she yawns contentedly, showing her snow-white teeth and the bright red of her curved tongue and throat, I marvel at her feline beauty.

'*Mau,*' I call again, imitating the wildcat summons to her kittens. '*Chirrr!*' she replies, stands up looking all intent, bounds off the bed over Moobli's dozing form and in a second has leaped onto the chair arm beside me and is thrusting her face into my beard. Purring loudly like a clattery motor, she hauls herself through my hands to rest both forepaws on my shoulder for a brief stroking. Then she looks outside as if suddenly aware she is missing good hunting, walks onto the window shelf and *maus* for her freedom. I open the window, she stands for a few moments on the sill, her eyes penetrating the darkness as mine cannot, then drops down to prowl the silent woods alone.

As I put the window back on its catch I remember the first time I saw her – on June 5th nearly two years ago, when I actually saw her being born. I had felt sure that evening when seeing her vainly trying to reach her mother's milk, the other larger wildcat kittens kicking instinctively against her, that she would never survive. Nothing could have been further from my mind at that moment than the thought that I would ever try to tame a wildcat. But I must now briefly return to the start of her story, to how the adventures with the wildcats first began.

Almost four years earlier I had acquired two wildcat kittens from Highlander Allan MacColl, who found them hissing and spitting in a deep ditch. Naming them Cleo and Patra I had reared them through that summer and autumn, not as pets but to go free in the wild. But fearing they might not survive the harsh winter after my 'soft' upbringing, I had offered them to London Zoo. At that time, however, the Zoo was making larger, more natural, areas for fewer animals and had a magnificent old 16-pound tom wildcat for which they wanted better accommodation. The upshot was that ferocious old Sylvesturr, whose instinctive hatred of humanity, increased by ten years of imprisonment, led him to launch furious frontal 'attacks' and spit and growl threateningly at the approach of any human, came to stay at Wildernesse.

My faint hope had been to try and mate this three-and-a-half foot bundle of incarnate wildness with one of my two females, who were only a little less obstreperous than he. That first winter Sylvesturr had escaped for nine days and when I trapped him he had contracted a near-fatal dose of pneumonia. I managed to nurse him through but as he convalesced in my heated workshop I noticed both Cleo and Patra were, at different times, spending hours on the window sill beside him in long intense gazing sessions that bespoke mutual adoration. Arranging a pen for the females next to Sylvesturr's main enclosure so they had access to him, I had hoped for the best. Cleo was the wilder of the two, slim, slinky and furtive, whereas Patra was the plump, greedy, opportunistic clown, and, as Patra spent most of her time in Sylvesturr's pen, constantly cavorting by his den door, I was sure he preferred her to her sister. To my surprise it was Cleo who produced two magnificent kits, Mia and Freddy, in late May.

When the kits were nine weeks old and well-weaned, I established them near a natural den in the four-acre west wood of firs, larches, ash and Scots pine. This spinney, with a 30 foot rock escarpment in its centre and small areas of swamp, tussocks and mossy boulders below the trees, edged an open verdant pasture in front of the cottage which stretched over a hundred yards to the east wood. Here grew huge gnarled oaks, with smaller ash, rowan, holly, birch, hazel and larch trees, leading to a grove of stately beeches which lined the banks of the main burn. I also turned Patra free. For five weeks, with the help of Moobli's fine nose, I tracked all the wildcats' movements through the two woods, finding out that Patra actually helped train her sister's kits to hunt and discovering too a great deal about wildcat behaviour. Mia, the shyer kitten, was the first to strike out on her own and had left her mother by mid-September. I live-trapped her again on October 4th but after that she never came back.

I had released Sylvesturr in early September, when the kits were well able to take care of themselves. As ferocious, resolute and unchanging in character as he had been since his Highland kittenhood, he went wild immediately. He established himself in a long wood a mile and a half to the west but his huge tracks and droppings showed he came back to the area roughly once a week for food I had set out. I live-trapped him on October 29th, found he was in fine condition and let him free again, that time for good. Occasional signs showed that he still came back to the area for food through that winter but far less frequently.

Freddy, however, had preferred to stay around his mother and often holed up with her in the woodshed, the home Cleo and Patra had used when kits. Patra too, after a long stay on a vole-plagued island up the loch, remained round the Wildernesse precincts, where I still fed them all.

But in late February and early March, the start of the true wildcat mating season, trouble occurred among the trio. Cleo and Patra had bitter yowling and clawing fights and, feeling it to be in her best interests, I finally released Patra on March 10th in a riverside wood near some rabbit-filled fields nine miles away, where a young wildcat believed to be a tom had also been seen! Wildcats are monogamous, faithful to a surviving mate for life, and as Cleo was then in restless oestrus, roaming the woods at night, I did not want Patra in the way of any possible new mating between Cleo and Sylvesturr.

My heart had leaped when rowing home in the moonlight on March 22nd to actually *see* the unmistakable form of Sylvesturr sitting on a rock near one of his day dens below the west wood. It seemed clear he *had* come back to try and mate with Cleo. The thought they might actually meet and mate again in the wild woods after almost a year apart was thrilling and I wanted nothing to prevent it. The problem then had been ten-months-old Freddy – for I certainly didn't want him injured or killed in a fight with his father.

In mid-March Cleo came into oestrus and was very active, rubbing herself over tufts, rolling on her back and crawling close to the ground with little growls.

When at the end of March I found Sylvesturr had taken meat from the west wood, leaving one of his vast tapering droppings nearby, I shut Cleo out of the pens but kept Freddy in. That night it had seemed an extraordinary coincidence when I received a letter from Edinburgh Zoo Park asking if I could provide them with a male wildcat – to try and mate with a female they already had there. Cleo stayed away four days but when she had returned on April 3rd anxious to be back inside the pens, I hoped it meant she *had* mated with her old paramour, and I let her in.

In late April, with the sun shining more strongly through the dispersing winter rainclouds, golden celandines and weakly nodding white wood anemones carpeting the floors of the woods in which birds had begun to nest, I decided to take Freddy down to the Zoo. I felt that apart from meeting a potential mate – which he was unlikely to do at Wildernesse – he might be an inspiration to youngsters who could become tomorrow's conservationists. It was agreed, too, that if he had not settled down happily in three weeks I could have him back. Before leaving, I had set out half-cooked and sterilized meats for Cleo in the woodshed and had turned her free again. I would be away almost a fortnight, for I had also to go to London on a long delayed work trip. Well, I had thought as she hared away to the west wood, she had the choice to go or stay – her future was up to her.

When Moobli and I returned in early May, spring was revealing its first beauty at Wildernesse. The first few bluebells were out in the front pasture, primroses imitated the sun palely in sheltered nooks, and new grasses sprouted where three years of sustained cutting had weakened the virulent bracken weed previously in almost total control. The larches were covered in light green tufts but while the oaks and ashes still lagged behind, the hazel bushes, rowans and birches were pushing out their first tips of varying jade.

I hauled the boat up on shore, waited for Moobli to catch up after his two-mile exercise run along the lochside, then walked up to check the woodshed for Cleo. There, at the top of the grassy path near the pens was a huge wildcat dropping. Sylvesturr had evidently taken advantage of our absence to investigate his old home. In the woodshed most of the meats had gone but of Cleo there was no sign.

Early next morning, the shed and pens still empty, I was wondering if I ought to try and live-trap Cleo, when something made me look out of the window – and there she was, walking along the pasture silhouetted against the loch waters, her striped tail curving low. I went out feeling sure she had heard the boat coming the day before, and *maued* to her. She stopped, looked as if unable to believe her eyes, *maued* back but then turned and walked back to the west wood.

Later we found her in the woodshed. She reacted fiercely when I appeared at the entrance, hissing and flaring then diving into one of the old den boxes on the logpile. I covered its entrance hole and carried her into the pens, and within hours she was her semi-tame self again, feeding avidly from fresh meat with little growls. She was very heavy round the gut and I felt doubly delighted – for not only was she almost certainly pregnant again by her ferocious and cussedly independent old lover, but with some twenty square miles of forested roadless wilderness to choose from, she had chosen to stay with us.

Now I was back, apart from normal wildlife studies, my main task was to locate an accessible golden eagle eyrie with chicks to observe and photograph. A pair of these magnificent birds whose wings, especially those of the larger females, can span up to eight feet, had spent the winter cruising the mountains around the cottage and had fascinated me. The mortality rate of red deer in my area had been high, so to help the eagles, which depend largely on carrion in the harsh months, I dragged several hefty carcasses from my woods up into the hills for them. As any stalker who has to drag stags or hinds down from the hills when they are shot knows, this is no easy task, so after heaving dead calves and hinds *up* to the three and four hundred foot levels, I began to feel I had earned the right to photograph the eagles at the eyrie, and also that my hard work might encourage the pair to use an old site not too far from the cottage. The golden eagle is one of some fifty rare Schedule One British birds who are rightly protected by law: it is illegal to disturb them at the eyrie during the breeding season, which lasts from early March to early August. But that year I had received the honour of being granted a government licence to observe and photograph them and naturally I was anxious now to do it successfully – which means without upsetting the eagles.

Before our brief trip south, I had already been on seventeen 'eagle' treks but had found only three eyries, none of which was being used this year. The pair which had wintered in my area, after feasting sumptuously in the dark early mornings on my last deer carcass, disappeared totally after mid-March, leaving the eyrie nearest me untouched.

I spent the next five weeks searching over a hundred square miles of mountains for a workable eagle eyrie, but also keeping a close watch on Cleo who had begun to eat and drink almost twice as much as usual and was lying up in her den most of the time. I was sure now that the early March oestrus and late May or early June birth was the normal wildcat reproductive pattern, for in mid- to late May when she was slowed down by pregnancy, there would be far more easy-to-catch surplus young birds, mice, voles and rabbits than at other times of the year.

My eagle treks, however, were proving a laborious, frustrating and dispiriting process. After one six-mile hike and climb in an area twelve miles from the cottage, I found an eyrie with two eggs. But it

Cleo eating a slow worm she'd caught. She had returned to the pens and, unknown to me at the time, she was some ten days pregnant.

was high in a V-shaped cliff fissure and the only place to erect a hide was in too-short heather on the clifftop itself, twenty feet above the nest, where I felt the risk of causing the birds to desert would be too great. An eight-mile foot slog over a circuitous deer path in the highest mountains south of my home revealed another eyrie with two eggs but high up on an almost sheer 800-foot cliff. Here the only place for a hide was on a scrub-covered precipitous ledge forty-five yards from the nest. Even with my 640 mm lens it was a little too far away for really good pictures. Well, it would have to suffice if I could find nothing better. Meanwhile, I kept searching – a traditional site described to me by a retired keeper was found after a two-day reconnaissance – but it had been destroyed by a rockfall at least two years earlier. Two more treks, six and eight milers, also proved fruitless.

On June 5th, hearing that an eagle pair had been seen several times on the far slopes of a high mountain to the west of my loch, I decided, despite having to wear brand new boots, to go and check the whole of its western face. Cleo had evidently spent some of the night in the west wood for on my way to the boat with Moobli I found the remains of a young missel thrush in a clearing in the west wood. Its feathers had not been clipped, as by a fox, but raked out in the wildcat manner. I wondered then for an awful moment if Cleo had in fact *had* her kittens somewhere in the wood and was just coming back to the pens for food, for that morning she looked slightly thinner. Either that or she had been suffering a false pregnancy, and the swelling was now going down. In any event, before I left on my trek I made sure the pen gate was tied open so she could go in and out at will.

It proved an exceptionally hard day on the hill as I had to search between the 800 and 2,000 foot levels and after covering eleven miles, mostly hiking and scrambling up and down the sides of a steep gorge, all I found was an old ledge site where the nest had been blown out by the wind. Even the discovery of a patch of rare yellow globe flowers failed to alleviate my depression as I boated home again with blistered feet. It was now rather late in the season to start erecting hides and working on golden eagles and I was beginning to admit defeat.

But before going indoors I took a look past the wooden door into Cleo's den where she had been resting up when I left.

An extraordinary sight met my eyes. What on earth was happening? It looked as if an albino mole had burrowed its way up into the den and was now resting on the surface hay, a grey-white, rounded object tinged with pink. But Cleo was licking it frantically while still lying on her side, as if the rest of her, which I could not see because of the door and its shadow, was somehow pinned down.

I raced indoors for my short-focus fieldglass and trained it through the den door again. She *had* given birth – to three fine kits who were suckling into the long buff fur of her belly, lined up like little furry piglets. And the object I had seen was the foetal sac of a fourth kit, which Cleo was now vigorously licking and pulling away from the feebly struggling form.

As I continued to watch silently, anxious not to frighten her, she swallowed the remains of the sac – her natural and instinctive response to keep the den clean and stop the scent reaching any possible predator – and kept licking the tiny kitten. It twittered like a little bird, waving its impossibly small limbs, and struggled unavailingly to reach its mother's teats.

As I prepared Cleo's evening meal and watered milk and put them through the wire enclosure into her bowls, I had no intention of disturbing the family that night but I said to myself: 'That poor little mite will never survive.'

Cleo was loath to leave her kits but she came out, ate half the meat very quickly, drank a little milk, then hurried back to the kits who during her short absence had been groping around blindly trying to suckle each other's warm bodies. Back in the cottage I poured myself a good stiff dram, and the first words I recorded in my diary were 'Here we go again. Cleo has stepped up her production one hundred per cent!'

Next day I hurried out and crawled through the gate, trying to avoid the new growing hogweeds as their broad leaves would provide shelter for the kits from the summer sun, and knelt in front of the wooden den door. Cleo, usually shy and fierce, had been astonishingly tame when she had given birth to Freddy and Mia last year and I expected she would be the same now, but to my surprise I was greeted by a low growling from inside the den. Wondering if she thought it something other than myself, I showed my face outside the entrance hole but she continued growling and even raised her lips in a flare.

The kits were smaller than last year's but similar for they had big feet, broad heads with low-set ears, were marked with chestnut and browny-grey stripes and elongated spots on a buff, almost cinnamon, background. Two were suckling but the others were lying on their backs in the crook of Cleo's rear legs, little shiny pink pads sticking upwards, and the kit furthest away from the teats was the smallest. Its sides looked pinched and I wondered if it had even had a suck yet. I would have to help this one somehow but Cleo was in no mood for any interference at that moment.

I crept out again, cut some small titbits of fresh meat and took them back. With these, the usual soothing talk and half an hour of patience, Cleo finally became half docile again and after stroking her gently she actually let me touch the kits. I knew now that if they were picked up they would emit loud piercing *meeoos* and Cleo would instantly revert to anger. Instead, I gently but quickly hoicked one of the bigger kits off a teat when it had stopped sucking, and put the smallest kit's mouth near the swollen pink nipple. Immediately it kicked like a TT rider trying to start his bike, thrusting hard, and got its hard front gums on the nipple. It then began to suck with such appetite that its whole body heaved like a suction pump. Oddly, Cleo seemed to feel little of this, as if her teats were not ultra-sensitive.

In the afternoon, with Cleo more trusting again, I managed to measure and sex the kits one by one, for if I was quick I could examine them before they woke up, realized they were in a human hand, and started squeaking for help. Sexing two-day-old kits is not easy for the amateur and mistakes can be made but it turned out later I was luckily right. The largest kit with the thickest stripes was a $7\frac{1}{2}$-inch female. The next two, both toms, were $7\frac{1}{4}$ inches each, while the small runty one, a female, was a mere $6\frac{3}{4}$ inches long. And she it was who let the side down and whistled out a '*Meeoo*' which made me hurriedly put her back near the teats as Cleo growled. But not before I had seen that she was the most perfectly marked kitten of them all – with a 'prettier' more cherubic face than the others, and the promise of at least five good dark rings on her tail.

What should I call them? All kinds of ludicrous names ran through my mind but as I saw them all lined up again, and Cleo turned her head upside down, splayed out her legs and lay on her back as if abandoning her entire belly with its swollen milk glands entirely to the kits, I settled for the most mundane names of all – Eeny, Meeny, Miny and Mo. Eeny was the big soft female, Meeny and Miny the two middle toms who looked alike, and Mo the weak little runt.

It seemed natural the kits should be smaller than last year's for there were four instead of two. Cats carry their babies in a uterus that is divided into two horns and it sometimes happens that a mother may have three kits up one horn with only one kit in the other. This could have happened in Cleo's case, with little Mo furthest up the horn as she was the last to be born. However, this was just surmise.

'She may be small but she's beautifully marked,' I quipped to myself, repeating a bachelor joke from the long-ago London years, as I backed out to join Moobli who had been standing outside the pen with furrowed brows.

But tiny Mo was very frail and during her first days I spent hours, gaining Cleo's confidence afresh each time, making sure Mo had a full chance to drink her fill, even if it meant slipping one of the other kits off a teat.

Once I went out to check the kits and received a nasty shock. The other three kits were lying asleep by their mother, only one with its mouth on a teat, but of little Mo there was no sign. Cleo growled when I went to reach over to feel the hay behind her recumbent form. Again I had to go to the cottage for meat but this time I dropped it more than a foot in front of her so Cleo had to stand to get it. She stood up, the suckling kit dropped off the teat and rolled over on its back, little feet flailing away as it tried to turn onto its front again, and Cleo stepped forward to reach the meat.

It was then I saw Mo, lying still at the back of the den. While Cleo busily chewed the meat I reached in and brought Mo forward. She felt cold, struggled weakly but was still alive. I warmed her in my hands and when Cleo settled down again I put her on a teat, pulling two other kits away about a foot so they would spend a minute or two groping their way back.

Wildcats are excellent mothers and will fight fearlessly against known foes or any threat to their young, but they follow the natural law of most predatory animals – they will not make any special effort to help preserve a sick or weakling youngster. As I went back into the cottage, I had little doubt that in the wild poor runty Mo would by now have been dead.

CHAPTER 2

Cleo with Miny, left, and Mo, right, after I'd brought the whole family into the cottage during a bad gale. Mo still needed help to reach her mother's milk.

The kits were only three days old when there occurred one of the most extraordinary incidents of all with the wildcats. Moobli and I were driving back to the boat from a supply trip and had gone well past the riverside wood where I had released Patra in early March, when a golden eagle passed overhead, leisurely beating towards some rock faces at the head of a small winding glen above the wood. I had always thought there might be an eyrie up there somewhere but I couldn't follow the eagle's flight because of the trees. I stopped the Land Rover, intending to walk through the woods and scan the faces with my field-glass – but we never got that far.

Suddenly Moobli put his nose down, zig-zagged about among the bushes and scrub cover of the wood, yawned noisily as he often did when starting out on a fresh wildcat scent and waiting for the order 'Track the pussy coots!' to go ahead. I gave the order and off he went. As I hastened in his wake I knew whatever he had scented could not be very far away as he was now moving urgently and tracking nose-high.

I had just hissed at him, the quietest signal I'd devised for making him go slower, as he was now just out of sight when I heard him making the little anguished sounds in his throat that meant he had something at bay. I burst through the trees to find he had put a wildcat up a large birch tree, and there, lengthways on a branch some fifteen feet above the ground, her striped tail switching from side to side, was Patra!

She was not spitting or snarling but her ears were laid down flat and she was making a low curdling growling sound of protest deep in her throat. I was sure it was Patra, not only because it looked like her but because Moobli had never yet treed a totally wild wildcat. As she straddled the branch, her big claws digging into both sides of it, and I made the usual soothing sounds I had always made to her and Cleo when Moobli had treed them in the past, I noticed she was extremely fat, her stomach bulging out over both sides of the branch. Surely, she was pregnant too!

Knowing she would stay up the tree for a while to make sure we had gone, I hurried Moobli back to the Land Rover, shut him in with one of his sterilized meat sausages which I kept in the vehicle for his meal if gales ever prevented us boating home, and went back to the birch tree. Patra was still there and as I talked soothingly to her I wondered if it was possible that she could have mated with Sylvesturr. She was almost two miles east of her release point, and Sylvesturr's main den seemed to be in the long wood a mile and a half west of Wildernesse. There would still have been five and a half miles between them, and I was sure wildcats are monogamous – certainly Patra had tried to entice the old curmudgeon to mate with her when they had all been in the pens together but had failed utterly – so it did not seem likely. On the other hand she could have mated with the young tom wildcat believed to be in her area. Well, there was one way to find out, for if she *had* mated with Sylvesturr the kits would resemble him, just as Cleo's did, and that was to try and catch her and bring her home.

I scattered just a few pieces of meat ostentatiously below her tree, for I didn't want her to have anything like a full meal, then boated home with Moobli for the box-cage trap I had used before for live-trapping the wildcats.

It was nearing dusk and some of the meat had gone when we returned, but I laid a small trail of tiny scraps then set the trap delicately with a juicy piece of red steak. We retired to a clearing over half a mile away, cooked food on the vehicle's little bottle gas cooker, and retired for the night. Next day the trap was empty. We busied ourselves checking dens and eyries in another area, then late in the afternoon I re-set the trap with fresh meat.

Just after dawn next morning we sneaked through the woods again. The trap door at the back of the trap was down and there, not spitting or dashing about in the cage part, was Patra, safely caught. She crouched in the wooden box portion, fat as a furry rugby ball, with a look more of reproach than anger in her black-pupilled eyes. Wildcats have excellent memories and I was sure she now recognized us. I covered the open cage part with a dark beach towel so she wouldn't rush at the netting, managed to carry her and the trap back to the Land Rover, and took her home in the boat.

Mindful of the fights when Cleo had attacked her in the pens in early spring, I clearly couldn't put Patra back with or even near her sister now she had kits. Luckily I still had half the fencing up in the woodshed where I had once partitioned off Sylvesturr, and in less than an hour Patra was safely installed there with a large hay-filled den box. I did not think Patra would mind a temporary loss of freedom for it was proving to be a cold wet summer and she was now relieved of having to hunt before and after giving birth.

Each evening now when I went to feed Cleo, I spent some time in the pens with the family, sitting on the grass, bitten to the point of torture by midges, to see if I could win the confidence of the kits before their eyes opened. I noticed Cleo now sometimes stood up to stretch and stayed high, as if consciously trying to make the kits stronger by forcing them to reach upwards for her teats. When she was eating, I removed one of the kits to just outside the den. Naturally it squealed *'Maoow'*. Cleo stopped eating, showed no belligerence, walked to the kit and after several delicate adjusting bites, picked it up not by the scruff of the neck as I'd thought usual, but by placing her bottom jaw right under its neck and very carefully carrying it back into the den.

On June 10th south-westerly gales began to rage around the cottage and I awoke to raucous bird calls. Some of the common gulls who had a breeding colony on an islet below the east wood had apparently decided their normal journey up the loch to feed among the weed beds at the far end would be too much hard work that day and were raiding the bird table. Two at a time they came flying over the pink and lilac blooms of the rhododendron bushes, landed on the table with flashes of grey and snow-white feathers, squabbled over the bread, then beat back to the shore with whole slices in their beaks. The tits, robins and finches who normally used it were all busy nesting in the woods now and finding plenty of insect and plant food, so the gulls were welcome to my scraps.

But the rain-filled gales were so strong and cold I was worried about draughts into the south-facing den, so I decided to pursue my 'friendship with the kits' policy by bringing them all into the cottage. First, wearing gloves in case Cleo turned nasty, I put the kits into deep hay in Cleo's old den box. She growled a little but immediately entered the box to carry them out again; I quickly shut her in and brought them into my study-bedroom. Now they were away from the noisy gales, Cleo soon settled down, and I had a fine session tape-recording her *'Brroo'* trills to them, and their squeaks when seeking her teats.

Runt or not, little Mo was the first to try and flare if I suddenly scratched the box. I noticed that when Cleo had had enough of them feeding she drew her upper rear leg right over her teats, thus closing the 'milk bar' to the kits. But they loved to be close to her when sleeping, stretching out both forelegs and little claws as far as they could across her body, their heads almost reaching her spine. Eeny, the biggest, was now $8\frac{1}{2}$ inches long. Mo still seemed to be constantly forced to the outside so through the day I occasionally pulled one or other of the kits off Cleo's teats and put her on.

As Cleo seemed quite tame again I decided to keep them all in overnight, maybe for several days if the gales kept up, and providing she used a litter box. When I fed her Cleo ate first but would not leave the den box for a drink until after dark. At 10.15 pm, having drunk their fill, I saw the kits were now lying around their mother, no longer in the nest of her legs but separately. Even though their eyes were not yet open the independent instinct seemed to have started to operate. Their pink pads were now turning grey before the final black.

It was hard to sleep that night because Cleo woke up just after 1 am and started to wander about, shifting through my paper files. She used the litter box all right, as I could tell from the unsavoury noises after the initial scraping of the debris. About ten minutes later she decided to drink the watered milk, lapping loudly, then a few minutes later to attack the dry biscuits, making such a racket as she crunched them up that sleep was impossible. Only the persisting gales and drenching rain prevented me from shoving them all back into the pens!

Next day, despite the weather, I had to battle up the loch for post and to keep an appointment with an excellent young keeper in a village to the south who had promised to show me an old traditional eagle eyrie. Unfortunately it had been blown out by gales and no new nest had been built. In the next week I would check the two eyries that should now have chicks and make a final decision about photography.

Returning home soaked and tired, I opened the study door and Cleo shot from under my desk into the den box with a horrible growl. It took me a while to soothe her down again. The sudden opening of the door had scared her, and like her previous kits and Patra, she had always been more wild when away from a fixed, known reference point. Inside the den or the pens she was reasonably tame, and inside the den box in the room too she would let me stroke her and the kits provided I moved slowly and gently, soothe-talking the while. But outside the pens or a few yards from the den box, and she was as wild as ever, ears back, growling like a dynamo. One had to be always on one's guard and not go beyond certain well-defined limits.

When I went to feed Patra later in the woodshed, I was greeted by a fierce spit and a foot-stamp right from the mouth of her den box! She had seldom acted this way towards me since she had been a kitten. I wondered if it was because she was angry at being incarcerated but of course there was no way I could tell her I intended releasing both her and Cleo and the kits later in the summer, that I was keeping and feeding them for their own good. Then I noticed Patra looked a great deal thinner and my heart started beating a little faster as I shone my torch past her into the entrance of her box. She, just like Cleo, had given birth to kits in my absence.

I could not see how many for they were deep in the box and I wasn't going to step through the partition and take a close look with her in that mood. I would leave it till the morrow. Back in the study I opened my mail. There was a letter from Roger Wheater, Director of the Scottish National Zoo Park at Edinburgh, thanking me officially for presenting them with Freddy. 'You will be pleased to learn I am sure that the Wildcat has settled down well in its new surroundings and it is hoped that you will come to the Zoo at some future date in order to view your presentation . . .' How wonderfully everything has worked out, I thought as I went to bed, happy to put up with Cleo's nocturnal noises and the occasional squeaks from the kits in my bedroom. With Cleo's two last year, four this, and now Patra's brood, we at Wildernesse seemed to be doing our share of putting a few more of these rare and beautiful creatures back into the wild – for Freddy was the only one I would give to a zoo. How lucky we had been to catch Patra again only days before she had given birth.

Next morning when I went out to try and check Patra's kits, the sad truth slowly dawned. I shone the torch past her grumbling form onto three kits, none as large as little Mo at birth. I could not believe it. No, it had to be a trick of the light coming through the window, shadows of hazel sticks atop the log pile. But as I continued to train the beam I knew it was nothing to do with light or shadow. Patra's kits had *thick* black stripes, some nearly half an inch wide, the dark marks on their legs were blotches, their heads were smaller than Cleo's kits, ears higher set, and their tails were also blotched and as thin as those of rats.

She had clearly mated not with Sylvesturr nor with any pure wildcat, but with some plain old domestic feline!

Interesting though it was to have shed some light on an old controversy – for it is believed in some quarters that wildcats will never mate with domestics when free in the wild – I felt most disappointed. Perhaps her early life around Wildernesse, Moobli and myself had made her less discriminating. For a brief moment I toyed with the idea of drowning all but one of her kits but I knew I had not the heart to actually do it. Really, I felt sorry for Patra as I recalled her general air of irresponsibility, her comical but vain attempts to arouse Sylvesturr's interest once he had mated with Cleo. Poor old girl, she had never really got anything right! Never mind, although my meat bill was now around £7 a week, I would feed them all and release them when they were ready. (Months later, the farmer's wife on the land near where I had found Patra told me that a pure black feral domestic tom had adopted the area in the early spring. But this tom was found dead in April after a fight with another large feral ginger male.)

By their ninth day Cleo's kits were crawling well but could not yet lift their bodies enough to walk. Their eyes were still closed but while little Mo was now a full one and a half inches shorter than the toms Meeny and Miny, and two inches shorter than the big female Eeny, I saw she was the first to use her little claws. Once I saw her actually rake some hay beneath herself. She was runty but not backward. The other three now lifted their heads and could spit audibly if they heard my approach but while Mo also lifted hers and had been the first to flare, she did not try to spit.

That night because the gales were still raging – at least my wind generator was working overtime to fill my desk light battery – and because my room was now a candidate for the strongest fresh air spray on the market, I transferred Cleo, kits and den box to the small rear room. The other reason I did not want to put them back outside yet was so I could still make sure Mo got her share of milk.

By June 15th, at the age of ten days, Eeny's eyes were two-thirds open, revealing the extraordinary bright china blue irises the kittens would have for about seven weeks before a greeny-grey tinge took over. In wildcats this turns to the final yellow-gold at about five months.

Next day, anxious to see if I could really start work on one of the only two possible eagle eyries, for I had found nothing up the glen where we had caught Patra, I went out early and drove to check the clifftop site twelve miles away. The nest was empty – the eggs had been stolen. That was the end of that. The gales had now died down but a misty drizzle had set in so when I arrived home in the boat I was not only tired but soaked through. But when I opened the rear room door to feed the wildcats I found a slight consolation. Although she was the runt of the litter, little Mo's eyes had started to open next, a little Chinese slit showing bright blue in the gap nearest the nose. Eeny's eyes were now fully open but not yet focusing. Before Cleo began her meal I saw her licking Mo's eyes to clear away the sticky secretions.

Again, when I first went in the other three kits lifted their heads towards me, flaring and spitting. But the little runt just lifted her head, trembling weakly on its neck, and seemed to be actually trying to peer, to see me. She was not only prettier than the other kits, she was cuter for she seemed to associate me with her mother and feeding. I wondered if perhaps she associated my soothing talk with the hand that had helped her so often to the teats. Certainly, from the purely human point of view she seemed more even-tempered than her sister and brothers.

CHAPTER 3

Having carried each of the other kits to the back of the den, Cleo comes back for the runt, Mo. She lifted each of them not by the scruff but by the whole neck, placing her bottom jaw carefully right underneath.

By the time the kits were two weeks old and the family had been in the cottage eight days, the powerful smell plus the fact that I would have to make a seven-mile boat journey to a sandy beach to fetch more cat litter, made me feel we might all benefit by the cats returning to the pens. I did not want to protect them *too* much as later in the summer they would have to start making their own way in the wild. They would need to learn the Highland rains were as much a part of summer as of the long grey winters. So when the drizzle ceased and the clouds cleared, giving way to blazing blue skies, I stuffed their den full of new hay, carried their box into the pens and opened it up.

Cleo immediately ran to the den, leaving all the kits where they were, so I took them out one by one, set them on the grass and withdrew. All the kits except little Mo began to *'Maow'* loudly. Cleo looked, her head going up and down, then out she came and very gingerly carried each one not back into the box but into the den, taking Mo last.

By now little differences in character and physique were emerging among the kits. Eeny, Meeny and Mo all had their eyes fully open but the second tom Miny only had a slight gap in his right eye. Eeny, Meeny and Miny now flared and spat loudly at any new approach of mine, and Miny's blindness didn't stop him from calling out the loudest when picked up or from trying to scratch and bite. He was undoubtedly going to be the fiercest kit. Among themselves, Eeny seemed the most lethargic and good-natured, her greater size making it easy to get to a teat whenever she wished. Mo, oddly, was now becoming the most active and once actually tried feebly to swat one of the other kits in play. All their foot pads were now turning black.

As I watched the kits I noticed that Moobli was looking up into the sky. I followed his gaze – to see a small flock of eight birds resembling woodpigeons flying to the west. But they were smaller, their wings made no noise at all and they had white tips to their tails. I had seen them about the area for the past few weeks and felt sure they were a small party of collared doves which have been busily colonizing Britain from central Europe since the early 1950s. They were the first I'd seen in the Highlands and as they eat much grain as well as weed seeds I wondered how they would fare, there being no grain crops in my area.

That afternoon as we walked through the west wood on a five-mile trek to double-check the rock faces above the lochside for a new eagle eyrie, a bird flew out of the trees above us with a clatter. Moobli immediately shot sideways – an odd safety device of his when anything surprises him suddenly – and I saw one of the dove pairs had made a nest in the fork of a small oak tree which had grown up in a near right-angled curve under a huge larch. I climbed up to the flat nest of twigs where I found two white eggs. Pleased, I hoped they'd rear their young successfully.

We found no new eagle eyries but on our return I went to check Patra's kits, for she was now much tamer than on the day of their birth. To my surprise the eyes of all three kits were opening at eight days, a good two and a half days earlier than those of Cleo's pure brood. But the blue was much duller, the eyes smaller with an opaque look about them, as if the kits were blind. Their calls of protest at being handled were much weaker than those of Cleo's kits, their mouths were dull pink instead of bright red, but it seemed they were not blind for when I held them close to the window they blinked, and I saw their pupils had formed normally. There is a belief among some naturalists that when wildcats breed with domestic cats the wildcat genes predominate but I now had the feeling (confirmed by the hybrids' development and behaviour over the next few weeks) that it is possibly the other way about – the wildcat strain is slowly and inexorably overcome by the domestic in successive generations. It was interesting that they developed a good deal faster than pure wildcat kittens.

Two days later, despite drenching rain showers, I drove to the southern glen and hiked over the soaking tussocks and heather with twin packs full of hide materials and camera gear, and up to the high shelf overlooking the other eagle eyrie with two eggs – my last hope. Just as before, there was no sign of any eagles and when I trained my fieldglass onto the eyrie the eggs were still there. It was now the last week of June; it seemed the eggs were addled and that the eagles had deserted. Wearily I slogged back down the valley, drove and boated home feeling finally defeated. But as I went to feed Cleo and her brood I noticed thankfully that Miny's eyes were now both open and the other three kits were starting to focus and actually look where they were going. None of them could walk properly yet but trundled along, bellies touching the hay, as they began to explore the den interior. And when I called to them I was delighted to see that Mo recognized the imitation *maus* and even took a few crawling steps in my direction.

Next day as I cut away choking bracken from the protective cages I had put round some of my young trees below the west wood to keep the deer from them, I noticed the collared dove sat tight on her eggs. Well, I could at least put up a hide and photograph *her* and her squabs – a far cry from eagles.

By June 26th the kits' two lower canines were all coming through and all but Mo tried to bite my fingers when I picked them up, a good hard nip which left the flesh white! Mo seemed by far the 'friendliest' kit in that her protests were less, and I spent more time caressing her than stroking the others. Miny was now the second largest, destined to be a big rangy tom, and he was certainly the strongest of the kits. The more I was finding out the more I realized one could not generalize about wildcats or conclude that females were shyer than toms, developed faster, or that toms needed their mothers longer. The individual variation was constantly surprising. When I went later with food, calling first as usual, Mo crawled to the curved hole in the den door and actually tried to climb over it towards me, thinking better of it when she saw the full four-inch drop on the other side! Two days later the kits' top canines were starting to break through and my enthusiasm for handling sessions waned considerably.

One afternoon I was watching them when a small head appeared from the bracken that covered the top of the den, a small tongue flickering – a very fat and pregnant slow worm was moving slowly over the roof and seemed about to fall down right in front of the door. As wildcats are extremely partial to these insect-eating, harmless, legless lizards, I removed her and let her go amid the riot of buttercups, blue speedwell flowers and delicate white-flowered pignut plants by the side of the cottage wall, where, until Sylvesturr had ended its life last year, another slow worm had once had its home. She could, I hoped, give birth to her live babies there and live in peace.

By the end of the month Eeny weighed $1\frac{1}{4}$ pounds, the middle two kits just over the pound while little Mo was still only 14 ounces. Yet she was nicely built as well as marked despite her size, for her head was even larger in proportion than tom Miny's. She was also the first I saw to swat out at the black tip of Cleo's tail – which wildcat mothers twitch about deliberately to help get their kits fast on their feet and to get their 'eye in' for later hunting. Mo also now cuffed out at her sister's and brothers' tails as well.

After cutting hay with a hand scythe much of one day, I took a small hide down to the west wood to set it up for the collared dove's nest. This time the mother bird did not fly from the eggs. Instead, I saw what looked like white down or little feathers just above the nest rim. Had they hatched out already? As I climbed up I noticed some bark and moss had been ripped from the curving trunk of the oak. There was only one egg in the nest now and the few wisps of soft white material I had seen were small downy feathers from the dove's belly plumage. As I looked more closely at the slightly disarrayed twigs and the scratch marks on the oak bark, there seemed only one explanation. Cleo and Patra were both enclosed with their kits – Sylvesturr had come on one of his regular hunting trips from the long wood to the west, and had taken the dove.

I climbed down again and reconstructed the whole scene in my mind's eye. He had perhaps seen the bird fly in at dusk, had waited in a thicket by the marshy area until dark, his great golden eyes blinking slowly as he had hatched his plan of attack. Then stalking slowly,

At three weeks, the kits lower canines were coming through, and even Mo could give a good hard nip which turned one's flesh white.

silently towards the tree, the old recidivist had gathered his powerful long back legs together for the kind of spectacular eight-foot leap I knew only too well he could perform, had sprung like a lightning tawny blur onto the trunk, made a couple of scraping scrabbles on the bark of the tree with his thick talons, and had seized the dozing bird before it had known what hit it. There were no signs of the dove's remains in the wood so he had probably run straight off with his prey to his home. What was most strange was the one missing egg. Had he come back for that on another night? If a crow had taken it later, surely it would have taken the second egg too – but it remained in the deserted nest until I took it myself some weeks later. The dove colony itself seemed to have settled near the farmland to the west.

By now, through a keeper friend, I had managed to locate an accessible eagle eyrie that contained a healthy seven-weeks-old chick in the 'black and white' stage, its dark brown wing, back and tail feathers contrasting starkly with the white down left on its chest, belly and head. We had erected a virtually invisible hide, not on a high ledge but actually on the ground under an overhanging rock on an opposite knoll some thirty-five yards away from the eyrie.

I will never forget July 1st that year. It was ten years to the day since I first left city life to start my wilderness life in Canada, and my first day in a golden eagle hide. It was like living in a magic, unreal paradise. For the first hour and a half I lay there, one eye peeping every few seconds through the camera aperture, the whole hide dark except for that living colour through the lens eyepiece, like looking into a huge and unique colour TV set. Without warning the eaglet stood up high and began to squeak, glaring into the sky.

There was a faint swishing noise then suddenly my lens was filled with huge wings as the mother eagle floated in, heat waves shimmering all round her, and, like a jet taxiing onto a runway, she landed on the eyrie with half a rabbit in her huge yellow talons. She landed with astonishing lightness for so large a bird, folded her pinions, then stood over her chick with a proud maternal air as it gripped the rabbit with its own talons and began to feed. An incredible moment, one of many over the next two weeks, and indeed the following years, as I worked to observe and photograph eagles. The main problem with this eyrie was the distance – a full twenty miles over the mountains as the eagle flies. For each return trip I had to boat thirteen miles, drive nearly sixty along the winding roads, and walk four just to get to and from the hide. It was often dark before I returned home so I had little time for any other work except caring for and studying the wildcats.

During the first days of July, with the kits walking unsteadily but with their bellies off the ground, I realized my belief that Mo was tamer towards me than the others because she recognized me as her benefactor was based on illusion. She was now showing a strong attachment to thin-faced Miny. And one afternoon I watched them playing together. They bit each other, rolling over and over, then sat up swatting each other's faces with paws like little ping pong bats with claws. But when I went closer to try and touch Mo she reacted like a firecracker, spat and shot away a few inches, flaring as furiously as did Miny, who had always been the wildest one. I withdrew my hand, talked soothingly, tried again – but with the same result. Now all the kits were flaring and even Cleo, as if affected by her kits' behaviour, had her ears back and was looking mean, though she wasn't growling. I left the pens.

Later I went back and found Mo was at the very back of the den, standing on her hind feet trying to push her way into a small crevice in its larch slab roof. When she realized from the reactions of the other kits there was something in front of the den she pulled back, saw me and again spat. I thought I knew why – until now she simply had not been able to *see* me clearly, her eyes unable to focus for more than a foot or two. Now she could see perfectly for several yards and while she had associated the *sounds* I made and the feel of my hands with being helped to the teats in her early days, the actual sudden looming *sight* of me scared her as much as it did the other kits.

Nevertheless I still had my own special feelings towards her and when, over the next two days, I found she was behaving like Mia, Cleo's female kit of last year, and trying to escape through to the rear rocks to be on her own between feeds, I became worried. This early manifestation of the independence instinct, which could later lead to a cunning marauding fox or even an eagle snatching an unprotected wildcat kitten, occurs with some kittens more than others. That it should now be developing first in the last-born runt of the litter seemed extraordinary. I spent the rest of the day blocking in the rear mesh of the pens with small mesh wire netting.

Cleo was now leaving the kits at dusk and *mauing* and prowling up and down, clearly wanting to leave the kits for a while and go hunting. But I could not let her range freely just yet because I had to spend a whole weekend away working on the eagles and did not want to risk the possibility of summer strangers boating up the loch and discovering what they might believe to be abandoned kittens. Instead, I brought them all into the rear room again, and put down two litter trays, and a supply of food and milk. Then I moved Patra and her hybrid kits into the pens for a spell outdoors. Anyone was welcome to one of *her* kits, I felt.

When I returned two evenings later Cleo was sleeping peacefully in her box with Eeny and Meeny stretched out by her stomach – but Miny and Mo had disappeared. Panic! I eventually found them, sleeping in a deep cardboard box in which I kept winter sweaters. It was the first time I had known any of the kits sleep away from their mother. As soon as I looked into the top of the box, Miny spat and flared, setting off Mo who did the same. And when I picked Miny out to put him back with his mother, he promptly sank his teeth into my thumb, then began to make little high pitched whirring growls, the first time I'd heard any of the kits growl. His bite was similar to accidentally grabbing a barbed wire fence. As for Mo, I just tipped the box over to face the den box so she could walk over when she liked. I fed Cleo and tried to make friends again but all the two day absence had achieved as far as the kits were concerned was to make them wilder.

Next day I tried an experiment. I wanted to see if a wildcat mother would accept or reject the kitten of another. Although individuals would vary, if one wildcat mother *would* foster another's kitten, it could be a useful factor in any long term breeding project, or if kits were found in the wild, as Cleo and Patra had been, but *before* they were weaned.

Patra was now as tame as she had been at the time of her release, and as she was now in the pens with her hybrid kits I determined to try her first, while keeping a close watch on the situation myself. As Mo was the smallest of Cleo's kits and was thus much the same size as Patra's, and as Cleo was after all Patra's sister, I caught Mo. At first she tried to bite but then went submissive in the hand once she was picked up, and I carried her into the pens. She immediately started to totter-walk into the den among her cousins, thick-striped and blotched, barging between them to seek a teat. I watched nervously, ready for quick action, but Patra showed absolutely no reaction at all!

It was as if she could not count or had known Mo all her young life – she didn't even sniff at the strange kitten. And Mo was soon guzzling away with two of the other hybrids. It was, I felt, an amazing as well as interesting result. I did not consider this brief introduction to be absolute proof of any theory about wildcat behaviour, and did not leave Mo in too long. Certainly I would never have left her with Patra overnight because I well knew the wildcat's ability to look unconcerned then suddenly attack another, without any change of expression or signalling of intent, as when Cleo and Patra had fought without apparent premeditation earlier in the year.

I tried it again next day when two friends came to visit. Once again Patra was completely friendly to little Mo. Not only that but mother and 'daughter' posed superbly just inside the den as cameras clicked away. When my friends had left I enticed Patra back into a box with a piece of meat, put the kittens in with my hands, for they were almost as tractable as domestic kittens, and transferred Patra and her family back to the woodshed. Cleo and hers could now go back into the pens.

By July 11th, at five weeks old, none of the kits was yet weaned but they all played and lay in the sun outside the den door with Cleo. The 'tail twitch' training had started in earnest now, with Cleo lying down in various places so the kits could chase the twitching black tip of her tail. This not only makes the kits surer on their feet but means the mother need expend little energy herself. Eeny still seemed the most advanced but was so good natured she often let the other kits bowl her over in play without protest. Miny was definitely the ugliest, with his thin face and hard wiry body, and he could now see the furthest. Whenever I approached it was he who flared first and darted to the back of the den with whirring growls, closely followed by Mo who still did not like my looming form, though she was the most submissive when picked up. Last year Freddy had been weaned at seven weeks old. If these kits followed the same pattern I would be able to release them in early August.

Two days later, during a dull and windless lull in the weather, with the biting deer flies, known as clegs, vying with the midges to make periods outside filled with energetic self-swatting, my ideas on releasing all the wildcats changed. In the morning I'd been watching the kits at play, delighted to see little Mo was now the fastest on her feet.

Mo (mid-July) is bowled over by the wiry young tom Miny – her favourite playmate.

Mo was fascinated by my typewriter.

She darted at the other kits, bowling them over backwards after playfully seizing them by the throat. In comparison, they still seemed rather tottery, making sudden dashes then falling over when they tried to stop, or looking about wildly after attacking heads of grass stems as if to make sure no enemy was witnessing their real weakness.

After transplanting fifty cabbages, I caught Mo and brought her into the house for a photo session, picturing her on the bed, the desk, scratching her ear with a hind paw while lying down, and to capture if I could the greeny-grey tinge that was now creeping outwards in her eyes, banishing the first bright blue. Oddly, she showed little fear of being away from the family for the first time. Then, with Moobli in the house, I set her down on the grass in the front pasture and retreated a few steps. She looked about nervously then started to run towards me making little *maus*, afraid to be out on her own in a strange huge world.

Delighted at such progress, I put her back with the others. Then I picked up Eeny who although the biggest was second only to Mo for being submissive in the hand. But when I set her down on the grass, she spat loudly and seemed most upset, so I put her back. She instantly ran to her mother for reassurance, to have a fuss made of her, and Cleo licked her all over vigorously. Mo watched all this, then went forward and started licking Eeny's tail! Then she licked her own paw and wiped her own face – twice.

At about 8.15 pm I fed Cleo, putting the meat and milk into her bowls through the netting as usual, then took the empty receptacles back into the cottage. When I got back, Cleo was eating – and so was Mo! She had been standing close by watching her mother chewing the meat, when she suddenly advanced to the bowl, thrust her head in beside her mother's – Cleo allowed this without even a slight growl – picked up a piece of meat, chewed it, dropped it, then picked it up again and ate as if she had been eating all her life.

So Mo was now weaned. Little Mo, the sickly last-born runt of the litter who I'd had to help to the teats or she would have died, had shown an exceptional advance and was weaning off onto meat before her bigger brothers and sister. What an eventful day.

After a supply trip next afternoon, I hauled the boat onto its wooden runners and found all the kits lying in line suckling Cleo in the grass outside the den. When they had had enough the three bigger kits went to sleep but Mo started to walk about below the big shady hogweeds. I caught her, brought her into the cottage and tried to feed her cat meat from a spoon. To my delight, she sniffed it then ate it all up as I held the spoon in my hand.

Wanting to know if she would actually play with me, I stood up to fetch a two-foot-long eagle primary wing feather with a broken tip that I kept in the camera shelves. But as soon as I moved off the bed, Mo flared in fear and ran to hide behind some sweaters lying near the wall where she growled like a tiny electric dynamo. I brought the feather back, talked soothingly to her with the usual calming words for a while, made sure she could see my hand coming towards her slowly, and managed to tickle her on the side of the head and under the chin gently with a finger. She looked nervous at first but liking the caress, calmed down again. I then showed her the tip of the feather with the other hand, making it dart gently about the bed just in front of her nose. Instantly she looked alert, ears pricked forward again, then she started swiping out at it, making great round-arm cuffs with her claws out, much faster than a domestic kitten.

It was at that moment, when I knew I was becoming really fond of little Mo, that I decided I would actually try and tame her completely.

CHAPTER 4

Mo alone on my bed for the first time. If I moved she darted behind sweaters to hide with high whirring growls.

Next morning, when she was exactly six weeks old, Mo's training began in earnest. After the usual making-friends session in the pens, I brought her in, set her on the bed and fed her a tiny piece of meat. I really wanted to see if she would actually *lap* milk yet. As she watched the honey jar lid filled with a strange white liquid come to rest on the yellow bedspread a few inches away, she backed up flaring slightly but then the smell hit her nose. Her head moving up and down slightly as she sniffed, she edged furtively forward and licked the outer edge of the lid. This produced nothing of the taste the smell had promised so she bit the lid. Immediately it tipped up, her nose went straight into the milk and at that precise milli-second she must have been breathing in for she sneezed violently twice, so hard that she tottered sideways slightly. But somehow a drop must have reached her mouth for she came forward again and was soon lapping away rapidly with the loud ticking sound Cleo and Patra had made when they were lapping milk as kits. *Eureka!*

When she had drunk enough, she started to explore the bed and the sweaters I leave on it as extra 'blankets' on cold nights. As I slid my hand near, fingers twitching provocatively, she did an odd little dance – ears back, eyes large, making a half circle backwards, and giving an occasional swift but feather-soft whack onto a finger with a clawless paw.

I knew that if I was to try and tame her to the *house*, she would have to accept Moobli who was now waiting anxiously but quietly outside the door wondering what on earth was going on. Warning him to be quiet and a 'good boy' I let him in. Mo immediately flared then reared as high as she could on her front legs, gave a loud spit and glared. But Moobli just kept still a few yards from her, and she seemed to realize that this huge animal she had often seen had never done her any harm before, and slowly subsided. After half an hour I put Mo back into the pens.

My plan was to spend more and more time alone with her each day, yet always return her to Cleo and her family afterwards, and also to keep Moobli in the act. Despite his gentle quiet nature, he was an exceptionally powerful Alsatian, measuring 6 feet $4\frac{1}{2}$ inches nose to tail tip. His inch-long canines were set in a massive bear-like head, and he could crack a red deer's thigh bone with one bite. I did not want to risk any feelings of jealousy on his part.

For much of the rest of the day I did some interior decorating, slashed down the weak fronds of bracken that were still trying to stage a revival in the front pasture, then at 8 pm brought Mo in again. She seemed slightly more amenable than before, playing with my moving hand, creeping towards my dark shadow, clawing and nibbling at my fingers. As I stroked her constantly and gently she actually started to purr. I let Moobli in again, but it was hardly necessary to tell him to stay still for he seemed as interested in her as I was. His expression was one of total fascination, kindness, even love. With ears cocked, he watched her from the foot of the bed not trying to touch or lick her, as if he also knew a young wildcat would have to be weaned to human and canine company too.

Although Mo often struggled, bit and scratched when in the hand I did not resist or treat her roughly but persisted with loving caress and soothing words. The only way such an animal could be tamed was by *constant* love, irrespective of what one received in return. The wildcat has bred right into its instincts over hundreds of years an innate distrust and hatred of its main persecutor, man, and the slightest attempt at defensive retaliation on my part would have aroused and heightened those instincts in Mo's tiny body. But here the love had to be expressed in *action*, not just attitude as I had practised with ferocious old Sylvesturr. Both Cleo and Patra had been fierce and mistrusting through kittenhood, only becoming semi-tame after months of patience. Cleo's second generation kits were only slightly less fierce than Mia and Freddy had been even though I had spent far more time with them, and the fact that Cleo accepted my close presence more often than not had clearly rubbed off most on little Mo, giving me a slight advantage in the process of taming her.

I was now letting Cleo run free some afternoons as she made it clear she wanted to leave the kits and go off to hunt. While she was away the kits bunched up together in the far corner of the den and never set foot outside until she returned. Although Cleo was fairly tame in the pens, she was quite wild if I came upon her outside. Her ears went back and she would dive for cover.

Next day I brought Mo in early. I thought she would have learned after yesterday's mishap with the milk lid, but again she bit it first, put her nose into it, then spilled some on her left foot. She sneezed, withdrew a few inches then shook her foot violently to get rid of the cold

Mo wasn't too happy out on her own. She walked about looking for cover, mauing at me to come and do something about it.

Left momentarily alone in the front pasture, little Mo maued plaintively after looking nervously about, then ran towards me. It was an important reaction in the progress of her training.

liquid. She then went forward and lapped most of it up normally. The odd sequel to this was that almost every time Mo approached milk to drink, she shook her foot hard first, a strange habit that has lasted over the years since. That morning I found she would, after eating and drinking her fill, nestle close to me if I was lying quietly on the bed. But if I was sitting she edged round to my back as if avoiding my eyes. I took her outside later so I could sunbathe on a camp bed, and, with Moobli sitting quietly several yards away, set Mo on the grass. She showed the same reaction, walking about looking for cover and every so often looking at me and *mauing* for me to do something about it. Finally she settled for the shadow under the camp bed. Moobli dearly wanted to pick her up or lick her and his tongue came out instinctively but when he once made a slight move towards her she flared and he stayed where he was.

After half an hour I put her back, wriggling through the mesh, into the pens. Mo looked at her sister and brothers – Cleo was away hunting – then back at me, *maued* and came right back struggling through the fencing again to reach me! It was confusing for her, me feeding and fussing her for an hour then putting her back, but she was not yet fully weaned from her mother's milk and I didn't want to take chances on too early a separation. But I felt it was another little victory. When Cleo returned at 6 pm she began swatting flying insects as the kits watched her. Meeny, who was the next kit to wean off onto meat, copied his mother and tried to swat a few too.

Two hours later I brought Mo indoors again – she could now climb up the tumbled blankets of the unmade bed and get on top under her own power, using her claws for body hauling for the first time. But when she wanted to come down again she was not strong enough to jump. She *maued* loudly. Then, as I did nothing to help, she half climbed, half slid down the blankets again, backwards. She searched under the bed, ranged the floor then settled for a brief sleep in a fold in the hanging blankets. It seemed their round-the-body touch was a substitute for the secure feeling of Cleo and the other kits' warm bodies.

When she had slept enough and started walking about again, I took her on the bed for a caressing session. She purred loudly like a tiny motor, rolled, kicked out at my hands with her feet, clutched them with her claws, bit my finger playfully and squirmed against the combined inertia of my body, arm and hand while upside down, her determined little jaw making her resemble a miniature lion. When I put her back in the pens and fed Cleo, I noticed big soft female Eeny was now also eating meat. It seemed odd that firecracker Miny, the fiercest, thinnest but strongest kit, should be the last to be weaned.

Over the next three days little Mo became more and more familiar with the study bedroom as her second home, she recognized the meat and milk dishes and ate and drank readily. But when I brought in the best den box filled with hay, which I hoped she would use for her little naps and as her own little bedroom when she grew older, she flatly refused to go near it. Instead, looking for a dark corner in which to sleep, she forced her way behind it and slept on the hard wooden floor shelves where I kept my battery record player, files and my boyhood birds' egg collection (before egg-taking was illegal). One afternoon she went to sleep with her backside on the cold concrete floor and her front paws on the first shelf. I left her for a while then lifted her gently

Although a runt, little Mo was the quickest to develop early skills. Here, she swipes out at a fly.

into the hay of the box which I had now covered with cloths to make it darker. She stayed for only a few minutes then came out and went into one of her playful little dances on the blue carpet.

These dances were crazy, a great feature of her kittenhood and delightfully entertaining to watch. She walked about the floor then suddenly splayed her front feet outwards, arched her back as high as she could, separated her toes as wide as possible with claws sheathed, sleeked her ears right back, wore a big-eyed surprised look, then went into a fantastically fast dance in which she shot forwards, sideways, then made her rear feet execute a hilarious circular waltz of their own. All the while her tail was upraised for some two inches above its base then dangled down in a lovely curve like that of a young girl's breast, with all the guard hairs fluffed up like a flue brush.

Occasionally she would glare at and suddenly caper towards a thrown matchbox or rolled acorn, pounce on it with high *'Eeya!'* shrieks, bite it, then bat it about the floor at a dizzying speed. Sud-

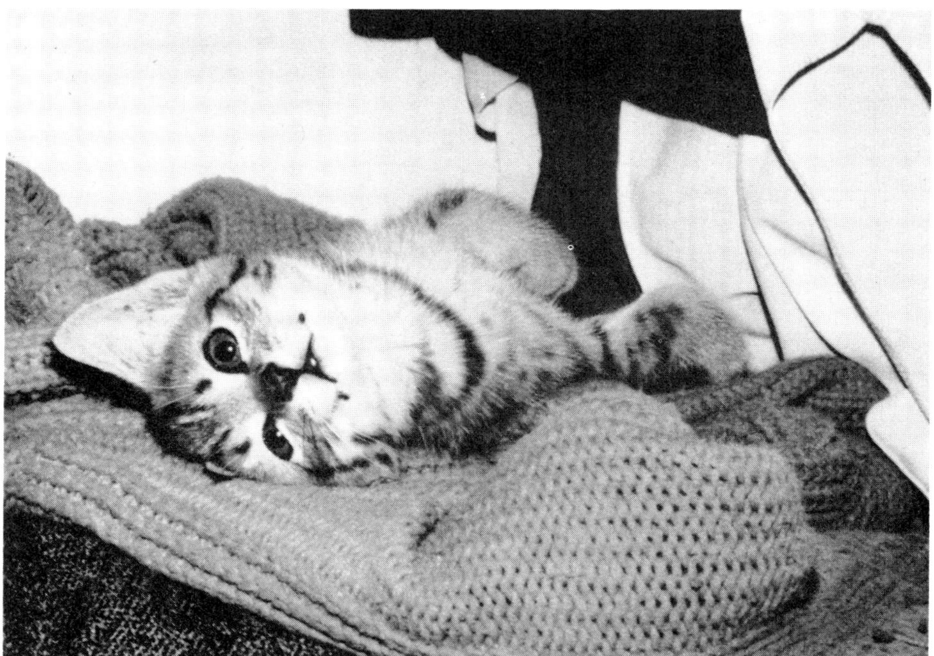

She was such a beautiful creature, I could no longer call her plain Mo. And so she became Liane. Lying on her back, she stared at me as if asking to be stroked, but if I moved, she quickly turned over with a slight flare.

denly she would break off and with a surprising leap land on top of the bed, hurtle round it, then leap off again, stalk towards me very slowly with an 'evil' look, then suddenly caper lightly away and take refuge under the bed or behind the den box.

So hilarious were these antics as I sat at my desk, that at times I found myself chortling silently, trying hard to stifle a loud laugh, for one sound from me would have cut them dead, but I found myself looking forward to them more and more. When a man lives alone in the wilds as I had done for most of the last ten years, genuinely amusing moments are few and far between. Now I was grateful to my little waif for bringing back the gift of laughter and as my affection grew I knew I could no longer go on calling her plain old Mo.

I suppose many single men have an image in their minds of their 'ideal' girl, an image seldom realized. I had met, won and lost mine, yet the mental picture, based also partly upon memory, persisted – she was a tall, shapely athlete with great green or blue eyes and long thick tawny hair, redolent of health and sun and sand and jungle. The name I had given this fading-dream woman was Liane. When I now saw my little tawny kitten, eyes flashing green-blue in the lamplight, performing her superb little dances so effortlessly, heard her shrieking *'Eeya!'* as she pounced with mock ferocity on her playthings, the two sounds and images came together in my mind. The very word 'Liane' could, when spoken in a high feline tone, sound very much like *'Mau',* the wildcat summons call. Thus little Mo became Liane from that day forth.

CHAPTER 5

Liane loved to doze after playtime stretched over Moobli's thick tail. The large Alsation grew to love Liane and bore her antics with great patience.

The eagle eyrie on the high ledge in the glen to the south which had contained two eggs in late June was still much on my mind, so on July 18th I boated out, picked up the Land Rover and drove down to see if I had been wrong about its desertion and if the eggs had finally hatched. I also wanted to photograph some badgers I had seen playing outside a sett in broad daylight when I had run out of film. It proved to be a wasted day. There was no sign of any eagles in the sky, the two eggs, certainly addled, were still there, and despite a four-hour watch in gloomy drizzle the badgers did not emerge once, and I finally boated home in the rainy dark.

I had left Liane in my room but, though she now usually ran to greet me with a little *mau* when I'd been out during the day, she was now terrified of the paraffin light coming through as I opened the door. She flared and shot behind the den box. She refused to eat or drink anything, so I quickly put her back into the pens, though I noticed Cleo was away out hunting.

Next morning I was glad I had put Liane back because Cleo returned with a dead vole. She deliberately showed it to the kits, then started batting it about with her paws between them as if encouraging them to be fierce and chase it too. When Meeny just picked it up with one clawed paw and looked at it, Cleo hooked it away and *made* the tom chase it properly. Cleo then ate the head as if showing them it was food, and allowed Eeny to drive her off it with high whirring growls.

This inducement to ferocity on Cleo's part was the prelude to the kits accompanying her on hunts themselves, teaching them to recognize rightful prey, and helping them learn to pounce correctly. Oddly the tom Miny, the most ferocious towards Moobli or myself, stayed out of all this. (He was not weaned until two days later.) This hunting training was essential to all the kits. Even if Liane did tame to the house successfully, I wanted her to hunt naturally too, so I could not separate her completely from the family just yet.

I left Liane with the others all next day, noting she was the first to eat and drink from Cleo's tall stainless steel bowl. This was quite a feat for she had to crane her head right over its four-inch-high edge and try to swallow with her throat pressed against it. None of the other kits were lapping milk yet. Later I was amused to see Liane suckling at Cleo's teats as she lay on her side in the grass, but unlike the other kits she kept her eyes open, cutely peeping through her mother's fur at my or Moobli's movements.

When the first sunny period came after a week of southerly gales and drizzle, I put Liane into the hay of the front pasture for some photos of her alone there. At first she crept through the herbiage away from me but when I had taken a few shots I called 'Liane' in a short abrupt squeaky voice. To my surprise she came bounding back – perhaps because I was the only thing she actually knew in the whole, new terrifying horizon.

When I put her back in the pens the other kits, who were in full play, gave her a boisterous welcome. With fat-faced stoic Eeny looking on, the two toms Meeny and Miny capered up to her and bowled her over with mock growls and for a minute or two I couldn't see which kit was which. As I tried to take pictures, Liane broke away, started to climb up the fencing peering out at me, then forced her way through the mesh and ran to where I stood leaving Cleo and the two brawling kits watching behind. What an astonishing breakthrough that seemed. Or maybe she felt the two toms were behaving a little too roughly towards her – runty last-born kit or not she was still a little lady – and realized she could find peace and gentleness with me.

Next morning I had her indoors again but after putting her on the bed I ignored her and did some writing. She dozed for half an hour then I suddenly heard loud *maows* and she was standing on the edge of the bed as if she had never climbed off it in her life, and was glaring at me wondering how to bridge the gap. Then she leaped to the floor, capered across and with a little jump onto the arm of my chair, hauled herself onto my lap and went to sleep. While it seemed I had finally won her trust, it made the physical act of typing impossible, for the jerky movements scared her – I had to resort to drafting my work with a biro.

Later, however, she blotted her copybook, or rather my bed, for the first time – squatting with great delicacy and blissfully half closed eyes she spread a large puddle over my bedspread. For some reason she refused to use the litter box of sawdust which Cleo had used when I'd first had the family in the house. I wondered if it was because she had been with me too much and was not receiving the correct toilet training wildcat kits have by watching their mother perform her ablutions then instinctively copying her as she scrapes debris over her wastes. That afternoon, Liane used the toilet area in the pens, and she did scrape grass and leaves over it. So again I felt I dare not separate her from Cleo just yet.

By now Patra's hybrid kits were more advanced than Cleo's although they were merely two thirds the size. I could not keep them cooped up in the partitioned shed so I opened up the whole shed for them. Once I had tried putting them in a separate smaller pen that ran crossways to Cleo's main enclosure; but, while I had proved Patra would accept one of Cleo's kits, Cleo simply would not accept the close proximity of her sister. And every time she saw Patra near the double mesh she launched an attack, trying to claw her way through it. This action upset all the kittens and made them bolt for their respective dens. The kits could have crawled through the mesh and might thus get hurt, so the experiment lasted less than a day. Back went Patra and her brood to the woodshed and down the loch I had to go for more sand and gravel for their litter areas, for they were now using half a sack a week between them. My meat bill was now higher than ever and, on her hunting trips, Cleo occasionally got in through the back door of the cottage to raid the workshop where I kept the sterilized meat sausages with which I bolstered all the animals' food. I was beginning to look forward to the time when I could release all the wildcats, though I still intended to keep Liane if her taming carried on its present course.

On July 22nd, with Cleo spending most of her days out in the woods, returning regularly around 6 pm to feed the kits, I brought Liane into the study again. She now had a new variation of what I had come to call her 'Zorba' dance. She still darted and shot about the floor with her paws twisted outwards but she now had a slower version too, where she swung her front legs out sideways like a fat girl running, and advanced and retreated in an odd minuet. At lunch time, noting the wiry tom Miny was inside the den box Patra had briefly used in the smaller pen, I shut him in and carried the box indoors and set it on the bed. I had been wondering if I ought to keep at least one of the toms as company for Liane, and, as she had always preferred Miny to Meeny, it seemed a good chance to find out how they got along alone. I removed the covering pine sheet – instantly Liane looked inside then started to lick Miny all over with great affection. He loved it, accepting the grooming with half closed eyes, but when I bent down to take a better look he flared at me and Liane hissed too.

I felt his cussedly independent company might make it harder to keep taming Liane and took him out again.

Around 9 pm Liane began *mauing* loudly as I worked at my desk and for the first time made a strange chirring noise in her throat, like a juvenile imitation of her mother's '*Brrrooo*' trill when greeting her kits. I thought she was missing the family but when I lay on the bed she jumped up, cuddled close purring, twisted on her back and pushed at my beard with her front paws. It seemed she had just felt neglected. Several times as I lay still she looked intently at my eyes blinking, then shot both paws out, her claws protruding slightly, as if trying to stop the movements of my eyelids. If this became a habit it could be dangerous when she was bigger, but a quick 'Na!' and a human hiss after the third time stopped her.

After her meal, I removed the shavings on top of the sawdust in her toilet box and replaced them with an inch of short hay clippings. To my relief she now used the box – fully. Later, after Moobli had come in after his gargantuan meal and was dozing stretched out on the floor, his huge paws sticking out straight like the cabriole legs of a piano, I noticed the room had become strangely quiet. Liane had found the warm living comfort of Moobli's great bushy tail, had put both paws over it with her head between them, and had gone to sleep! He was well aware of this of course and was looking at me sideways, the whites of his eyes showing with the strain of peering at me in that fashion without moving his head and disturbing *his* little friend!

Moobli, after my constant reminders, now knew exactly what was required to keep Liane trusting him and that any sudden movement from him or me scared her. He often kept dead still as she walked about what had once been just his and my room, nervously sniffing up at his gigantic form. I could not let this extraordinary moment go by without record; I sneaked to the shelves and managed to take several photos of Liane draped over his tail. It was to become a regular dozing place for the little kitten over the next few evenings. Moobli never once showed any jealousy and behaved like an angel.

When I took the kit outside again Cleo was back from hunting, a dead woodmouse on the pen floor, and she greeted Liane with a welcoming *'Brroo'*.

On July 23rd, with a low mist hanging over the hills and loch, I boated out for supplies then went to explore the rocky cairns and woods that ran for almost a mile along the single track road near the sea where Cleo and Patra had been found as kittens two years earlier. I was looking for the den of a huge old tom wildcat I had seen two years before that, and which had eluded me on two previous searches. That afternoon I found it – between huge boulders below a cliff and to the right of the entrance a dry mossy rock held the greening bones and skulls of rabbits and small birds. It did not appear to have been used recently. As I watched some rabbits hopping in the fields towards the sea, a friend drove by. He told me the big old tom had not been seen for at least a year, but he had seen a new but smaller wildcat twice in his car headlights, and when he had seen it in early spring it had appeared to have bald patches along its flanks. I recalled that old Sylvesturr had sustained so heavy a moult during one April that his skin had shown pink along his lower flanks. It seemed possible this new wildcat was also a tom. Hmm, I thought, when it came to releasing Cleo and her brood maybe this rabbit-filled area where she herself had come from would be a good place. There was a good disused den and the new wildcat did not sound too large so her tom kits would probably be safe enough.

When I returned home at dusk, however, Cleo and her kits, including Liane, had all gone! Had she taken them off into the wild already? In panic, I raced round her hunting area in the west wood with a torch, ordering Moobli to 'Track the pussy coots!' He tried his best but the frequent leg cocking on tufts and scent he was finding on mossy rocks told me all he was finding were recent fox trails. Only later, when I went to fetch his biscuit meal and nightly meat sausage from the rear workshop, did I find the answer.

Cleo had carried each one of the kits into the workshop, the door of which I'd not only failed to lock but had foolishly left open, and had installed them on the piles of sacks I used for bird hides near the free food supply! She reacted as usual when not in the pens, flared and hissed, backed into a corner with ears flat, then shot past me and vanished into the night. I had a hard task rounding up the kits into a den box but I finally got all four back into the pens and left the gate open, sure Cleo would shortly return to them and the food and milk I set out. Sure enough, she was back in half an hour and I closed the gate in case she tried to spirit them off somewhere else during the night.

As I came back with the paraffin lamp, remembering I hadn't yet fed Patra and her three kits in the woodshed, there was a sudden flurry in the grass by the shed wall. And as Moobli shot forward instinctively a small striped shape shot across the path, its little claws scrabbling on the earth as it forced its way through a tiny hole at ground level and back inside. One of Patra's kits! They too had the wanderlust upon them and while Patra could not get out, it seemed her kits were now investigating the world outside on their own. I blocked the hole up.

It was now clear that Cleo was ready to take the kits on their first forays away from the den, and in a week or two this would lead to them going with her on proper hunting trips. And Patra's family were obviously in the same situation. I now had to think carefully about all their futures.

CHAPTER 6

Liane appropriated my desk swivel chair as her favourite bed. She would not give it up without a struggle. Often she leaped onto the arm and forced me to move up to make room for her.

In the early morning three days later an extraordinary event happened which made me hasten my plans. I had just dressed, sitting on the bed, when I caught a glimpse of a large bird floating silently past the window barely thirty yards above the front pasture. Hoping it might be an eagle, not just a buzzard, and knowing both birds circle round when hunting, I screwed the big telephoto lens onto my camera and closing the door on Moobli who was whining oddly, I stole slowly outside to try for a shot. Sneaking to the west corner of the cottage, I was just in time to see the bird, a large female buzzard, sailing serenely over the lochside edge of the west wood. But when I looked over to the wildcat pens an astounding sight met my eyes.

There, in the smaller pen, partially obscured by the brambles, stood Cleo, her tail as thick as a bottle, ears back and growling in a way I'd never heard before. She was standing high, ready to attack. And facing her, just outside the pens and glaring with its big orange eyes as if trying to mesmerize her, was the red-brown form of a large fox. It was visibly wilting before the ferocious wildcat stare and very slowly was starting to back up.

I knew I was far too close with the big lens to try for a photo but my main thought, although the kits were clearly safe in the den, was to get that fox out of there, and teach it a lesson it would not forget. I drew back slowly, then tiptoed swiftly to the door, and as Moobli, who had sensed something odd was happening, shot past me I hissed, 'Go on boy. See him off!'

By the time I reached the corner again both fox and Moobli had gone, racing past the rear of the cottage eastwards to the main burn and its waterfalls. Moobli had an astonishing turn of speed when he needed it. I ran as hard as I could, stumbling through the long bracken on the north hill, hampered by the camera and long lens, then I heard Moobli's deep baying bark in the east wood. Cursing my mere human speed, I reached the burn – just in time to see Moobli floundering out of the far side of a pool, the fox just a foot in front of him, and Moobli's jaws actually clamp onto the very tip of the fox's thick tail. But, as Moobli struggled to get his footing out of the pool, the fox, now on the edge rocks, tore itself free, bounded forward and made a tremendous leap of at least ten feet onto an almost sheer slab of rock, hooked its forepaws over the top and vanished into the herbiage.

Moobli, whining, the fox's tail hairs still sticking to his mouth, danced about, knew he couldn't make such a leap from the standing position himself, and was about to rush downstream to cross the water flow where the far bank was lower and go after the fox when I called him off. I knew he hadn't a chance, for the long legged fox can run as fast as a hare and, knowing the ground and its trails intimately, would in that quarter minute have been several hundred yards away through the thick bracken and summer growth. I found later that in those few trembling seconds I had tried to take a photo with the impossibly long-focused lens – a fuzzy out of focus shot it turned out to be too, of the middle of the slab of rock. It is at moments like these the erstwhile wildlife photographer experiences the heights of frustration – if I'd had a standard lens on the camera, I could have moved faster and photoed the incredible moment when Moobli's jaws had clamped onto the fox's tail. Most impressive, apart from his speed for so large an Alsatian, was how he had sensed while inside the house that something was wrong, and how in a trice he had picked up the running fox's scent.

The fox had obviously come sneaking around in the hope of picking off one of the kits, possibly Patra's who had been straying through that little hole for several nights. I did not think it would return during the daytime again after such an experience, but it was probably cunning enough to know Moobli was normally kept in the cottage at night, and might just return in the dark a few days later for another attempt.

While I would shortly have to release all the wildcats so they could learn to hunt well with their mothers, I knew Cleo and her kits were not yet *quite* ready to go. She had proved herself fearless when confronting the fox and I felt reasonably certain she could protect her kits from it. But what if it returned with a mate? I could no longer leave the pen gate open for her to hunt alone in case the fox, or foxes, got to the kits while she was absent, or she herself took the kits away. I had two other reasons for not releasing them in my own woods – if Sylvesturr was still around, though I'd seen no signs of him since March, he might just possibly attack the tom kits; also there would then be too many wildcats in my rabbit-less area for a natural balance of wildlife. No, she and her kits, and Patra's, would just have to be kept penned in a few days longer.

I now made my final decision about their release. I would keep little Liane but Cleo and Eeny, Meeny and Miny would be set free amid the rocky cairns and woods near the rabbit-filled fields by the sea. And Patra and her kits would be set at liberty in a large conifer wood on the far side of the loch where there were many voles, mice and other prey. I would leave a food supply for both families each week until it ceased to be taken.

The immediate problem was Liane because, being the smallest kit, she could still wriggle through the mesh of the enclosure, so she was in more danger from the fox than the others. The answer was both simple and expedient. She was now well used to the study-bedroom after her daytime visits, so from now on it would be her permanent home. I, too, would virtually have to *live* with the young wildcat for many weeks until she was large enough to look after herself, but if all went well the process of taming her would be enhanced. If she needed the companionship of her mother and brothers and sister, as well as to be with them for hunting training on dead voles or mice that I would now have to trap, I could, until they were finally released, let her spend daytime periods in the pens while I kept frequent watch. But every time I went on a trek or supply trip she would stay in the room.

Having worked this out, I caught Liane and brought her indoors. She was very playful for about an hour, then *chirred* loudly and leaped onto my lap and went to sleep. Carefully I transferred her to the haybox and did not hear a sound from her all night.

I was woken up by a sudden thump on the bed and strange pokes all over my body. Liane had leaped up and was investigating the strange humps and bumps now below her on the normally flat bed. Heart beating fast, I didn't want to scare her off, yet wasn't sure what she was going to do – my face was exposed – I kept still and soon found out. She crept right up to my face, watched my eyes blinking and once again shot out a paw to stop the movement. *'Na!'* I hissed and she leaped two feet backwards, then started one of her Zorba dances over the bedspread, advancing and retreating, ears back, looking ferocious. I swizzled my hands about under the bedclothes and she watched every movement, leaping and chasing after them with swift jerky movements like a big squirrel. She 'caught' my fingers with both sets of claws and bit down into the clothes – and I was thankful for their covering.

Presently she tired of this and, apparently hungry, began to nuzzle round my face and beard, smacking her lips – the noise wildcat kittens make when they want some food. Then she forced herself down past my chin into the bed itself as if looking for a dark place she could explore. I wiggled my toes, she felt where the movement was coming from and like a mole burrowed down to them. The next thing I knew she was biting my toes! I soon stopped moving them and started twiddling my fingers instead. I felt her forcing her way up towards the movement when suddenly I felt a sharp excruciating pain as, en route to my fingers, she encountered something else of interest, a naked hairless mouse no doubt, and bit me in a most embarrassing place! For years I had slept naked summer and winter, did not even own a set of pyjamas, but if this bed burrowing became a regular early morning feature – and it did – if one did not wish to surrender, well, one's wedding equipment to a wildcat, one would need to make other arrangements. From then on I kept my underpants on in bed. On this first occasion, however, I managed to grit my teeth and stay still, finally enticing her back up to my neck area with frantic finger twiddling of my left hand.

Before long she emerged by my face, looked over at the window, leaped down and with a fast sideways tail-dangling caper, shot across the carpet, up onto my chair, and found a new resting place on my work files on the window shelf. Here, allowing me to get up and dress, she dozed for about half an hour, then started grooming herself with long licks up and down her shoulders and forelegs.

When she started playing about on the floor again I put down my large round shaving mirror, magnifying side outwards, to see what would happen. At my movement she had shot with a hiss to the top of my clothes case under the bed and from this perch she sniffed with bobbing head the strange new shiny object on the carpet. Then, overcome by curiosity, she jumped down, walked round to the side as if not wanting to stalk it directly, then came in front of the mirror. The sudden appearance of a huge wildcat kitten's face twice the size of her own made her flare and leap about two feet backwards. But as nothing happened she again crept forward, saw the face begin to appear again, ducked back, but each time she grew bolder until she was staring at the strange cat full in the face and sniffed it. Most peculiar – no scent! Within four minutes she was reaching behind the mirror with her claws for the other big kitten and was clearly mystified there was nothing there. But within ten more minutes she had somehow worked out that the other kitten's movements corresponded exactly with hers, that it was therefore something only to do with *herself*, and she never played with the mirror image again. Not even on the normal side where the image was the same size as she was.

After she had eaten and drunk, still shaking her left paw violently before lapping the milk, she scratched at the floor, looked round, spied her sawdust and hay-filled litter box and went to perform her toilet, her tail pumping up and down like a pump handle. That completed, she stood and with her left foot delicately raked debris over her leavings. It seemed at that moment she had been 'toilet trained' by Cleo for she used the box and covered her wastes this way for the next two days. I was delighted – prematurely as it turned out.

By now the early blue of her eyes had been banished to the outer edges and the replacement greeny-grey was gaining a lighter gold tinge. The few spots mingling with the stripes down her sides were elongating into the final shorter stripes of her adult coat, and the tram-lines of dashes that swelled out and in again from the dark brown line down her back were also starting to merge together. While still a runt compared with the others, she was fast, well built for her size and seemed to be thriving on the fare I gave her.

On two fine but sunless days, I took Liane outside, set her down in various places in the hay of the front meadow, then backed quickly away. Each time she looked furtively about then ran to me *mauing* loudly, hopping over the long grasses with four-footed leaps for forty yards or so like a rabbit. When I took her to the shore she was terrified of the tiny lapping waves and instead of running to me again, she cut up towards the house, walking slowly through the long bracken below the fence with plaintive *maus*. It was amazing how her stripes mingled with the stems and shadows so I lost sight of her and had to call in Moobli, who brushed through the bracken and found her in seconds. When I picked her up, although his noise had scared her, she was not standing at bay or growling but just crouching down looking at him quietly. She knew by now he was a friend.

I carried her up to the pens and dropped her through the camera hole after removing one of the wire strands. She ran straight into the den and Cleo greeted her with a murmured *'Brroo'* not even getting up, as if the kit had never left her side. I was sure Cleo couldn't count. Liane then barged in among the other kits who also showed no surprise. Although one kit was suckling away and I watched for a quarter hour, Liane made no attempt to get milk. I was sure now that I could separate Liane totally without problems, and at night I brought her back into the cottage again.

One morning Moobli and I went on a short trek to a badger sett about a mile away that had been unoccupied since the spring of the previous year. As we walked the deer track below the steep wooded hills, we found badger signs aplenty – little scraped toilets with fresh deposits here and there as they had rooted about during the nights for worms and grubs – but when we reached the sett it was still empty. Searching in a wide circle around the sett, we had just negotiated a huge rockfall above it when Moobli picked up scent which brought us to the edge of a steep dry gorge. To my surprise there were five-toed badger tracks there, going upwards along small sloping ledges, at times so sheer we were almost climbing. Yet it seemed the badgers used the trail often for it was well churned up and I felt sure we'd find a new sett around the four hundred foot mark. We worked our way up to around 800 feet where the slope was shallow again, but on the hard packed short heather surfaces the trail just seemed to peter out. Baffled, we returned home.

The far corner of my study was a mass of white where all my files had been ripped open and tossed about, as if by a whirlwind, sheets of paper were lying everywhere with very few left on the shelves. The mischievous cause of all this rumpus, who had clearly had a fine old time on her own, was lying neatly curled upon what was *now* her favourite bed – my desk chair. 'You can't discipline a wildcat,' I repeated to myself as the lovely little brute greeted me with a *'chirr'* and stretched herself lazily. Well, I'd feed the other cats before I tried to restore order.

When I went to the woodshed, I found that the bottom half of the double door had somehow come open and Patra and her entire brood had vanished. Moobli found them in seconds – now *they* were ensconced in the workshop and the kits dashed everywhere. Patra snarled, spat and dashed through the door and into the north hill bracken but I forbade Moobli to go after her. She had broken into the carton of sterilized meat and between them they had devoured two whole sausages and had used one of the sacks as a communal loo. As I rounded up the swelling kits and put them back into the woodshed, leaving the *top* door open so Patra could join them later in the dark, I knew the time for her final release was nigh.

That evening it took me a full three hours to sort the right papers into the correct files. The wanton little miss, meanwhile, full of beans after her meal, discovered that a sheet of paper lying on the carpet made a jolly good toboggan if one leaped on it at speed and braced all four feet. And when, finally, I sat down to type out my diary I found she was not prepared to give up her new bed to her huge foster parent, not without a struggle that is. She leaped onto the arm and forced her way down beside me, making me shift up into a cramped position to make room for her! There she stayed stubbornly and went to sleep, only the hot lump by my backside reminding me now and again that she was still there.

The only time she stirred was when I'd just come to the end of eating my supper, so thinking she wanted to get down I shifted over slightly to let her out. Instead, she climbed up on the desk beside me, watched me put the last forkful in my mouth, then as I licked my lips swatted out at my tongue. Ouch! None of this did I really mind for the little wildcat was turning out the most affectionate – and demanding – animal I had ever had. But I had a suspicion that Moobli, who was dozing on the floor, was taking a rather dim view of all this, for I occasionally caught him, with head up, looking at me – was that an expression of contempt in his eyes?

Around dawn next morning, after playing moles under the bedclothes again, I heard Liane land with a little thud on the floor and gratefully turned over for another fifty winks. No way – suddenly there was a great deal of swishing. She was at the files again and as I looked up she had, quite by accident, worked a thin hard cover file so that it rested halfway over a small wooden divider on the shelves. It had become a see-saw! She walked up to one end and down it went, to the other and down went that end. She did this several times, enjoying herself thoroughly, until the papers suddenly slithered out onto the floor – and the great crash sent her scurrying for shelter on the suitcase under the bed.

She now began to assert herself in other odd little ways. If stroked while asleep she woke with a loud squirmy note of protest. But if I left her alone too long, especially at night, she walked about the floor *mauing* loudly to be picked up for a cuddle, then she purred and lip-smacked in my ear. If I picked her up when asleep she gave an angry *'maow'* and within a minute or so, as if deliberately, would leap on the bed for a pee. I wasn't having that so the first time she did it I seized her gently but firmly by the neck scruff, pushed her nose into it lightly and thrust her into the haybox. This manoeuvre worked – for a time. I soon learned that her peaks of greatest activity, when she indulged herself in file shuffling, Zorba dancing, sledging along on paper sheets, swiping out and biting at the tassels on the dangling bedspread, were at 10 pm, 1 am and 7 am. Unfortunately, they did not coincide with mine but, when one lives by choice with a young lady of independent mind, one needs to make a few adjustments.

Patra, in fact, did not return until the evening after we had found her in the workshop, so she had left her kits alone for nearly twenty-four hours. They were certainly ready to go free and I felt Patra would turn vicious towards her kits if closed up again in the woodshed.

On the last day of July, I enticed them with meat into a den box, slammed the door shut by tugging a long cord through an eyehook, and in blazing sunshine took them over to the long conifer wood on the far side of the loch by boat. Patra growled most of the way over, pushing her claws through the ventilation holes of the box.

With a parcel of sterilized meat sausages in my pack, I carried the box of cats up a steep bank, negotiating about half a mile of difficult twiggy going between the small spruces and tussocks until we reached the first green open patch. Then setting the box in a thicket, spreading the meat around so she could see it, I opened the door, said 'Goodbye old girl' then took off quickly back to the boat. I felt if I had stayed looking at them I might have changed my mind.

I felt little emotion as we boated home as I intended to check up and renew their meat supply on fishing trips, and would probably see them again. But Moobli lay in the well of the boat looking miserable.

I cleaned out the entire woodshed floor and the litter area – good fertilizer for the fruit trees – and then went to see what new havoc little bossy boots had created in my study. But as I peeped into the window, the place seemed in perfect order and there she was, lying all alone and still on the green cushion of my chair, her big ears twitching slightly, pink nose between her big paws, little sides heaving as she breathed. It was almost as if she were keeping my chair warm for me. It was bad enough that the normal term 'pussy cats' had been perverted by myself to a fond 'pussy coots' when referring to Cleo and Patra, but as I looked through the window at little Liane, she looked so utterly sweet lying there a new term coined itself in my mind – 'Pussy Queets'. Ludicrous perhaps, but it seemed apt, and I'm afraid to say from then on I called her Pussy Queets more often than Liane throughout her kittenhood. At times it degenerated even further – to Pussy Tweets or Pussy Poots!

From her warm cramped perch beside me, she now launched into new kinds of mischief. She took to leaping on the desk and chasing the moving typewriter carriage, flicking at the ends of the paper, swatting my typing fingers. Once she seemed to leap from nowhere and land her front quarters right into the well of the machine as I was in full spate so that two of the hefty sharp keys hit her. She squawked, leaped out and shook her paw in pain – and never tried that particular stunt again. Instead she took to Zorba dancing on the desk itself and, after one intricate dervish-like double pirouette, swiped my tin of biros and pencils right off and into the fireplace. She also started burrowing *through* the middles of the loose files, like a huge shrew going berserk and hurling up debris from a woodland floor. She used her claws more often too, usually to hook off paper clips. Once she hooked the photograph of a woman friend out this way, slung it up into the air, batted at it with her clawed paws, as broad and spiky as huge black sundew flowers, and waltzing about on her rear feet only, whapped down upon it time and again with her front feet, then blithely chewed it to shreds.

Yet she was a fastidious eater like her mother, not greedy like her auntie Patra had been, and only ate what she needed. She often left choice meat and milk in her bowls for later snacks. At night she still spent time sleeping while draped over Moobli's bushy tail – while he dared not move for fear of upsetting her. Then she would wake, look up at me, then leap and haul herself up the old coat on the back of my chair and force her way down beside me. Once I forgot she was there, sat back too far, and she squawked loudly *'Mauw'* like a duck grabbed by a fox, as her foot had been momentarily trapped. I still took her into the front meadow for walks. Usually she ran to me but sometimes she made for the nearest cover of small hazel, birch and oak trees amid the long grasses. But if it was drizzling or Moobli went too near, she hared straight back into the house and study. This was a useful reaction for it meant she now regarded the study as home and I hoped she would continue to do so in mature life.

Before I finally released Cleo and the other three kits, I wanted to make absolutely sure Liane no longer needed her mother and to clinch this process I resorted to a low-down cunning trick. I put her into the pens for a short period each afternoon, gladly noting that while she walked in among the bigger kits as if she'd never been away, often out-wrestling them all in play, she never once tried to get suck from Cleo. But when I fed them then I put in only the roughest of foods: stodgy meat sausage and plain water in the bowls. The other cats would eat and drink but Liane, used to the best, sniffed and turned up her nose at the rough fare. I then removed her again and fed her the choicest meats and milk in the study but returned to the pens to feed the other cats good food and milk too. Liane didn't know this, of course, and I think she came to realize that only in the study would she receive the fare she liked most, and not with her real family. It was sneaky but effective.

One night I returned from a fishing trip with Moobli to find Cleo and the kits had disappeared. Moobli soon traced them. Cleo had forced her way through a loose wire in the camera hole and had carried each of the kits into the newly vacated woodshed, her home when she herself had been a kit. Eventually I got them all back into the pens. It was time for them to go free too, for it was now essential the kits learn to hunt in the wilds with their mother.

Next day, as I had to go out later on a post trip and to collect sawdust for Liane's litter box from a local sawmill, I went out to the pens wearing leather gauntlets, now necessary with Cleo's kits, and got Eeny and Meeny into the big den box easily enough. Cleo followed them in to eat the bait meat, but Miny hid in a rock cleft at the back of the den. I managed to haul him out a little by the root of the tail and seized his neck scruff as he bawled and tried to bite, then thrust him into the box with the others. We boated and drove to the site I had selected, the rocky cairns in the woodlands by the rabbit fields, and I

For a wildcat, Liane sometimes had a sweet, almost 'pretty' expression.

staggered up with the box to the disused wildcat den – she could use it if she wished or find another.

I set out the meat sausage supply where she could see it, opened the front door flap of the box then retired some twenty-five yards away with Moobli to watch. At first Cleo stayed crouched, her kits all behind her with saucer eyes, looking fearfully around, then, bit by bit, eased herself out. She put her feet on top of the box and with eyes huge and black-pupilled looked cautiously at the strange new area. She knew it was goodbye this time, she knew by the distance the boat and Land Rover had travelled that I meant it now, there was no going back.

She then went back into the box among her wide-eyed kits as if she didn't want to leave. I walked forward again, picked her up with soothing words and gently set her on a rock below the den after kissing her neck. Then she went, sneaking low, and was out of sight in seconds. Twenty yards away she started calling to her kits *'Maoo, maoo'* and they answered with little *maus* and one by one filed over the rocks and through the thick grasses. My last look was of Miny's backside before they disappeared for ever.

'Goodbye Cleo,' I said, feeling a terrible lump in my throat. 'God bless and look after you, dear old lady.' And that was that.

As I closed the box and carried it back down, Moobli looked as if he could not believe what I was doing and made a move forward with his head low as if he wanted to track them as of old, but I called him back. If I felt sad as we drove away and boated home, Moobli seemed even more downcast. He lay in the boat, forcing himself to sleep as he always did when unhappy, his eyes red and bloodshot when I woke him up by trying to explain 'It's for their own good, old son.' How he had loved running out ahead of me each morning to check *his* charges, his playmates, were still there.

Although I set fresh sterilized meats out for Patra and Cleo and their kits on my next few fishing and supply trips, Cleo's remained untouched after two weeks and Patra's after three, and I brought her den box home again. We never saw either wildcat again.

For more than two years the lives of Cleo and Patra had been intertwined with ours. Now all we had left was little Liane.

CHAPTER 7

Although she could climb slanting logs quite well, it took a bit longer before Liane, at nine weeks old, learned to descend backwards under her own power.

For the whole of August Liane lived in the room with me, apart from accompanied outdoor excursions, and as she grew stronger, weighing over three pounds by the middle of the month, there were times when 'little Liane' was as much as I could cope with. The very night we returned from releasing her mother she seemed to realize her family were no longer outside as usual, that she was now on her own with us and she was suddenly more wary and kept her distance. I had to learn that a deep underlying love was not in itself sufficient when dealing with such an animal that is wild and independent by nature – that love had to be constantly communicated in every movement and action. I'm convinced wildcats are 'psychic' and somehow know when the status quo they have been used to is about to change or has altered, even should there be no outward sign.

I knew *she* knew her mother had gone and this knowledge made *me* nervous. When I picked her up and was stroking her as I had done many times before, she suddenly and for no apparent reason yowled noisily, scratched, then sank her teeth hard into my right forefinger. I reacted instinctively, protectively, without thought, and threw her away. She hit the floor, rolled over, then spitting and growling with a high whirring sound, dashed for refuge on the suitcase under the bed. There she crouched hidden by the 'curtain' of the tasselled bedspread, and when I lifted it up trying to make friends again she spat and slammed down her foot against me, just as her father Sylvesturr had done throughout his life at the approach of any human.

At that moment I thought I had undone all the work of taming her and I cursed my stupid reaction. I sucked and washed my gashed finger, then put some meat and milk by the side of the case. But for two hours she just stayed there, refusing to touch it. I realized then what a great part mental telepathy played in trying to tame a wildcat – any fear or nervousness on your part communicates itself to them and upsets them, overcoming even the love one really feels. I knew better than to try and make any more approaches, left Moobli in the kitchen where he normally slept, and just stayed working at my desk, not daring to go to bed in case the noise above frightened her even more.

At about 2 am I heard a movement – she had sneaked down and was lapping her milk. I watched sneakily from the corner of my eye as she ate some of the meat too, then walked a few steps towards me, apparently thought better of it, then leaped onto her second best bed, a pile of sweaters on an old sea chest near my bed. When she was asleep I tiptoed quietly over and slid between my own blankets.

Next morning I woke to find her still there. Pretending not to even look at her, doing without breakfast so I wouldn't have to open and close the door, I moved quietly to my desk for some typing – a noise she was well used to. Suddenly there was a little *'mau'*, a thump on the back of the swivel chair, and she was beside me, trying to force her way down between me and the arm. Apart from shifting up to make room I again made no move to touch her. The next thing I knew she was clawing at my right arm as I typed. I kept typing with that arm while I sneaked a finger of my left hand over and darted it about on my sleeve. She immediately started to claw it, but in play now. By just ignoring her it seemed I had won her back.

During the next few days her play sessions became even more intricate. She tried to climb the hard mahogany leg of the desk, but kept slipping because her claws could not penetrate far enough. She knocked my plastic, slide viewing lens off the desk and began batting it about the floor – it was the first time she had chased a large hard object this way. She seemed far faster than a domestic kitten, her movements delicate but deadly accurate. She lay down, tail flicking, stalked it, then dashed in a high romping run like a tiny lion. To replace Cleo's black tail tip, I made another plaything for her, a hard rolled ball from a bright green and red stiff magazine cover and suspended it on nylon fishing line from a string near her litter box.

She played with this for hours, dancing, swatting, flirting, waiting for it to swing overhead as she ducked low, then whacking out at it after a swift turn, her eyes huge, intent and black. Sometimes she dived on it with both sets of claws, pinning it down to the floor, then doing a forward roll with both feet and her teeth fixed into it.

At night with the paraffin lamp hissing away, I noticed her looking at moving shadows on the white walls. So I made strange animal objects, pecking birds, fluttering butterflies with finger shadows and she leaped onto the bed and began to chase them. But on the second night she worked out they were not 'real', began to look from them to my hands – far faster than it had taken Moobli to work this out when he was a pup.

Of course, the dozing, occasionally snoring form of Moobli proved too much of a temptation for her. Naturally he was becoming slightly jealous of her, at being ordered to 'sit' every time he nosed towards her out of interest, and he never really liked it when she snoozed over his tail. So when I saw her dancing deliberately near him when he was lying in the *lion couchant* position with his head up watching her, I was a little worried – one chop from those mighty jaws and inch-long canines would have ended her nine lives in a trice. But when she sleeked her ears back, did a rearwards waltz, then danced in like a mini Muhammad Ali and delivered a quick left and right with clawless paws to his huge fawn muzzle, he merely blinked and drew his head back half an inch. Again she went in, right, left, right, and away, but he actually seemed to be *enjoying* it. His mouth suddenly dropped open in a huge grin, his long pink tongue lolled out and he started panting as he often does in a warm room. But when Liane saw the great mouth and teeth she did a reflex leap right over his head and landed on the bed. Not for long. Moobli saw me looking at him approvingly and started wagging his great tail. This sight again proved too much for Liane and down she jumped, dived straight onto it, clutched, bit, rolled over, then started kicking at it with her hind feet. Now Moobli didn't think much of this treatment and he whined *sotto voce*, but he took it and let her get away with it. For the next two weeks it became a regular feature of Liane's nightly play – if Cleo's tail tip represented a vole then Moobli's was a big hare – but never once did the long-suffering dog snap at her.

Another game she invented with him was to suddenly throw herself on her back right under his nose, lie there with her legs splayed out, then whack out with each paw at his chin. If his nose went too close, she flared – while still lying on her back on the floor! One night, however, Moobli decided to cut this latest caper out – he whoofed right in her ear. At that close range it must have sounded worse than a thunderclap – and she was on the suitcase under the bed so fast he never saw her go. I felt it essential I stay out of these dog and wildcat play sessions, for it was important they worked out their own adjustments and came to trust each other.

Every time the urge came upon her now, seldom more than once a day, Liane used the sawdust haybox, never once performing her toilet anywhere else in the room. It seemed she was perfectly toilet trained for which I was grateful. For a time her droppings were too liquid, but I solved this by mixing some crushed dog meal with her cat meat for one of her three feeds. She not only scraped hay over her wastes now but once or twice also tried to rake debris over her left-over food too, a real wildcat trait.

She also used her suitcase refuge if I walked into the room wearing my rustly oilskins, or when I made a wood fire. She stared at the flickering flames nervously, showing no signs of wanting to investigate or sit near on the mat like a domestic cat. Other differences I noticed were that she seldom erected her tail hard when stroked, and never once 'kneaded dough' with her front paws when on my lap.

One morning I woke to feel a strange weight by my side nearest the wall, then heard a loud purring. Liane had apparently spent part of the night sleeping outside the covers next to me. Then she walked right over my face, sniffing my nose and eyes with the cold tip of her own then right along my hair line, making lip-smacking noises as she did when searching her mother's belly fur for a teat. She would now lie upside down on the bed and push against the mighty human hand with all four feet, and with considerable strength, taking harmless tiny bites at my fingers. When she shot out a paw onto my eyelids now I didn't hiss but nipped one of her rear legs from under the clothes, changing the type of play, or else made sure my eyes were shut when she looked about to do it. It seemed she now trusted me completely again. That day she reached up against the fish box shelves in which I keep all my natural history files, and dug in her claws for the muscle strengthening or 'claw sharpening' movements for the first time. She was two months and five days old.

One day, feeling it would do her good to have a long spell in the fresh air, I put her into the pens alone for three hours while I wrote indoors. When I went out again she set up a great racket, *mauing* loudly and trying to claw her way out through the mesh. I went round the back, opened the gate and she shot into my arms, *mauing*, purring, sniffing at my nose and eyes and gazing upwards into my face as if to be sure it was really me after such an age! When I brought her indoors the fuss continued, though, again, she never rubbed herself against me like a domestic cat. Later she sat on the desk as I typed, gazing at my face so intently I began to feel embarrassed. Once she shot out a paw and one claw stuck lightly in my lip. I just prised it off and gave her a hard look!

By mid-August, although she was still rather small for a wildcat, just one foot eight inches long at thirteen weeks of age, she was strong enough to inflict considerable damage if she really wanted to. Her Zorba dance sessions now had an air of ferocity about them. The initial ears-back coy look as she drifted and waltzed about with turned-out toes, was now supplanted by a really nasty black-eyed look as she dived onto my hand, clawed hard with both sets and pulled her head down for a bite, which was sometimes quite hard. Then she would dance away again, sideways, looking back archly as if to say 'Now now. How DARE you!' – as if *I* had attacked her! Then in she would come again, pounce with a loud *'Eeya!'* clutch, bite and kick, and be away again in a trice. It was lovely to watch, painful to experience (though she only seldom drew blood) and I fervently hoped she would not do it when mature. I found at this time that she *would* accept a light tap on her backside, providing it was in play and no harder than she would have received from her mother, without reverting to ferocity. This usually stopped her biting too hard. But she was full of confidence now, in her own den. In fact she was just about taking over the joint!

Although she could now climb the logs that were propped up against the outer wall quite well, it was not until August 9th, when I refused to lift her down again, that she learned how to descend under her own power. Her instinct was to go down frontwards but not being a squirrel with reversible rear feet, she couldn't. So she tried half sideways, slipped a bit, clung on backwards without moving, sideways and slipped again, and finally completed the manoeuvre by going backwards completely, having difficulty prising each set of claws clear as she went down. But she learned quickly – next day she did it correctly without a second's thought. Unfortunately Moobli came round the side of the house at that moment. The sudden sight of him moving towards her outside scared her, so she ran into the thick bracken to the east. I could not find her so sent Moobli to scent her out – suddenly she was streaking for the room, her den, and safety, which was just the reaction I wanted. Some days when everything had gone well, I could actually call her out *from* the cottage and over eighty yards of the front meadow to me, but only on fine dry days. She hated wind, didn't mind light rain, but if the vegetation was really soaked she walked along reluctantly, shaking her feet free of moisture every few steps.

One night when chasing her hefty cardboard ball she must have

Mo, the runt, alone in the hay in front of the den where she was born. At two weeks' old, her eyes are just open and she is prettier and more gentle-tempered compared to her brothers and sister, Eeny, Meeny and Miny.

I brought Mo, now five weeks' old, into the cottage for handling session. She was less upset than the others at being away from the family, but was still nervous and spat at any sudden movements. The bright china blue of her eyes is now being edged out by a greeny-grey which will eventually change to the pale gold of an adult.

When put out in the open on her own, Mo would run back to me as if afraid of the huge strange world. As I lay quietly she climbed onto my boot for a better view.

From the time she was weaned, Mo spent more time inside with me. I did not separate her from the family completely for she still needed Cleo's milk and training.

Once, after a boisterous fight with the other kits, Liane broke away and climbed up the pen fencing to get through to me. I felt it was a great step forward in our relationship.

Out in the sun, Liane often panted like a dog, flanks heaving, showing her bright red tongue and throat. Her eyes are now their final gold.

When I bowled an apple towards her, she chased it, cupped it in her paws and bit into it . . . the taste did not suit her palate.

At nineteen months, with the thick, darker coat of winter, her runty childhood behind her, Liane has become a sinuous and powerful young wildcat. She still comes through the window some nights for reassuring contact, and even sleeps beside me on the bed covers. Here, although she looks fierce, she is merely yawning.

tripped, for she fell right over into her food, getting cat meat and meal into her fur. She had to be cleaned up like a naughty child. To be accurate, much of the time it was like having a recalcitrant teenager on my hands. As her litter box was now a little small for her I changed it for a large plastic baby bath I'd once found on a dump and had kept because it looked 'useful', half filling it with sawdust and the hay covering she preferred. She took to it like a charm but to my chagrin not as a toilet. Instead she cuffed clouds of hay and sawdust everywhere and used it to *lie* in, elbows hooked over the sides, as she swatted out at her swinging ball. I noticed now that her tapetum, the screens at the back of an animal's eye which in the dark reflect helpful light back onto what it is looking at, were now flashing in the lamp beams with a bright blue-green light.

Once when she was playing this way she started to make terrible yowling screeches, not kitten noises at all but like a full grown cat being strangled. At first I thought it was a new 'play' cry of ferocity but she was rolling about on the floor, apparently in pain. She had swatted the nylon line above her ball too hard and it had swung round twice, wrapping itself tightly round her foot. And the harder she pulled to get away the more it hurt. When I tried to help she yowled louder, spat and tried to bite me. But I managed to cut the line and free her.

Although she seemed to be over it quickly, jumping onto her chair bed beside me, she slept far longer than usual that night and ate like a canary next day, and even looked thinner. She sneezed twice too, which was worrying. For all her boisterous play sessions I had to remember she was still a runt and delicate for a wildcat. In the wild I was sure she would never have reached the weaning stage.

The incident taught me never to leave her alone in the room too long and when I went out on treks or supply trips, I made sure the line was well out of her reach. It would be awful if it ever caught round her neck.

CHAPTER 8

Liane's playfulness was developing an air of real ferocity. Here she clutches and bites at my sweater, kicking out with her back feet as if to disembowel her 'victim'.

Trouble seldom comes in a single package. Two days later when Liane had fully recovered her spirits, I was scything away at the long hay below the cottage, the sun beating down and a fine stiff breeze keeping away the midges, when there was a sudden light tawny flash, a loud *'mau'* and Liane was right by my feet. The blade of the scythe had missed her by inches. I had not seen her coming and although I'd left the doors open I had no reason to think she would venture outside of her own accord. The only times she had done so before was after I had been calling her patiently for about ten minutes, really working at it, with Moobli sitting behind me.

A few evenings later I felt she was not having enough *hard* exercise in the room and wanted to make sure her claws and leg muscles were developing the sort of strength they would have needed in the wild. So at dusk when her eyes were at their best, I put her on the trunk of a long slim ash tree south-east of the cottage and clapped my hands lightly, hoping she would climb up a little way. Moobli heard the clap, thought it was a signal to him, and came dashing up. The sight of the fast-moving Alsatian apparently coming for *her* galvanized her into action and she shot up the trunk like a squirrel. I called her but she kept climbing higher and higher, then she looked down.

The height she was at further scared her and she looked upward again, spied a small branch at about thirty feet and with slipping claws, obviously tired, she gained its safety. Then she looked down at us, a lost tiny little figure, even more terrified at the unaccustomed height, and began to *mau* plaintively.

At first I thought I could call her back, for she had managed to climb down the seven-foot larch logs upended in the porch. They, however, had been slanting and the perpendicular smooth trunk of the ash tree was a different proposition. I called and called and she walked back and forth a few paces on the branch, still crying out, looked as if she would try and back down, then funked it. I had no ladder at that time, just a triangular heavy scaffold device I'd made for painting my roof, but there was no way I could prop that up against a vertical tree trunk. After several minutes, she looked as if she might just try and jump down, so I decided to fetch a blanket in the hope I could drape it over my outstretched arms and so catch her if she did.

I turned, had only gone two steps when I heard a rush of air and a horrible thwacking sound. She had jumped, apparently making for me, but because I was moving had missed and instead of landing on the soft earth, the front part of her body had struck a big rock only thinly covered with moss. She yowled, kicked, fell over while instinctively trying to get onto her feet, then lay on her side. As I stared in disbelief her guts seemed to be swelling up. In awful remorse, I picked her up, stroking, kissing, talking to her and carried her back into the room.

She was not breathing and I thought she was dead but suddenly her sides began to heave fast as breath began to be sucked again into her winded lungs. But she was stunned, out cold, one eye totally shut, her mouth open. I felt she must have a fractured skull, and would still die. I laid her on some soft sweaters and nursed her for an hour. I felt her little body all over gently and could find no obviously broken bones, but the swollen stomach area area made me fear that she had suffered internal damage, though I could see no blood in her mouth or anus.

In the morning I woke early to find she was still alive and could just about walk but she was weak, tottery, and her left eye was swollen shut. She was very subdued all day and could not eat till nightfall. She was wary of me as if feeling it was all my fault – as to a large extent it was. The following dawn I woke feeling stifled – she was draped across my neck. And when she began to purr again and sniffed my nose and eyes, smacking her lips looking for milk, I quickly fetched her some. In three days her eye was back to normal and she resumed usual play and her Zorba dances, but the ferocity and shrieking attacks on my hands ceased. In a way, as far as taming her was concerned, the fall seemed to have helped.

By the third week in August she was putting on weight again, becoming quite hefty and looking more like a wildcat. I now let her have free run of the whole area, and she soon leaped through the window and discovered the woodshed for the first time. Sometimes I fed her in the pens so that she would remain used to being in them, should I need to leave her there for any reason later, or if she became fierce. Occasionally, if she had not seen me for a few hours, she would flare or spit at my sudden approach. I could never forget that underneath the apparently successful taming so far, she was still a wildcat. I developed a habit of always scratching gently at the ground before

Perched on the twigs of the bird table (never used by the birds in summer) Liane appears to be biting her fingers with nerves at the height.

sliding her bowls towards her, a gesture of continued friendliness. Moobli continued making his own adjustments to her out of doors, and sometimes I found him sitting staring at her intently, brows furrowed, tongue lolling in a grin as she sported about, though she was never far from the gap between the shed and the cottage wall.

But every night Liane slept in my room, usually leaping onto my sweatered back from the floor and making the chirring note of greeting, demanding to be petted. Yet she never clutched at my bare hands or neck skin, as if knowing her claws would hurt. Sometimes she insisted on lying along the crook of my arm, so preventing my typing, or pushed her head between my arm and side, purring with a clattery sound like her mother. Often she licked my fingers or face frantically with her eyes closed. She was now more affectionate than any domestic cat I'd known, had more variation in her play, and was far more demanding. Outside in the sun she often panted like a dog, her flanks heaving, mouth open showing her bright red tongue and throat. The interim greeny-grey of her eyes was now almost entirely replaced by the final gold that would grow paler with age.

Towards the end of the month Liane developed digestive trouble; perhaps she had eaten something outside that didn't agree with her.

Her droppings were very loose, and she refused to eat or drink. She began to lose weight rapidly. Wondering if the close atmosphere of my room, my pipe smoking – though I cut it down when she was indoors – the playing of music and civilized indoor life, were upsetting her, I put her den box in the kitchen with Moobli. I set it on the window seat behind the long refectory table I had made from birch planks. She had now lived five weeks in my room and perhaps she needed more time in a fresher atmosphere. She was now big enough for her toilets to make a strong impression on one's nostrils. I knew Moobli would not hurt her now, he was as fond of her as I was, and I left the doors open so she could go out, as could he, whenever she wished.

But this did not work either. She sometimes slept thirteen hours at a stretch, ate almost nothing, and began to urinate on the carpet and look even thinner. I knew she had to eat or drink *something* so using a vet's hypodermic syringe without the needle, I forced some milk with a pinch of aureomycin and half a powdered cat condition tablet down her throat twice a day. By the 29th she was still loose so I added a little kaolin and chlorodyne – the great foreign travel antidote to runny human stomachs – into the syringe mixture as well.

I now had another worry regarding her. I needed to go for an urgent short work visit to London, to complete some biological research and meet two publishers. Understandably, knowing the animal's popular reputation, none of my local acquaintances wanted to look after a wildcat. Besides, all cats are better left on their own territory and I felt no-one would boat down to Wildernesse just to set out her food. She would just have to come with us but not when she was in this condition. She was also bad-tempered and bit my hands several times.

Luckily, the new 'medicine' seemed to do the trick. On the 30th, obviously feeling more sprightly, she chose to go into the pens of her own accord for a few hours. That night I set a piece of half cooked steak on my desk chair as a little treat but she only ate half of it. As Moobli was staring at her eating, drooling slightly at the smell of the steak, I said, 'Go on boy. Eat your food.' He sneaked up to the chair and very delicately with the tips of his front teeth slid the steak away. Liane's expression should have been photographed. He'd never taken *her* food before. She leaped up, flared and went to clout his muzzle, but he was away onto his bed in the kitchen with the meat. The next piece I gave her she ate very quickly with little growls as she chewed away with her side carnassials.

Later on she leaped on me purring loudly and I caressed and stroked her after the three difficult days. She seemed fatter already. (Her morphological balance was most delicate and seemingly she could lose or gain weight in a matter of hours.) She then jumped onto the bed, watched the battery record player's turntable going round for a minute or so – with Giuseppe di Stefano's magnificent tenor filling the room with '*Che Gelida Manina*' – then she leaped right on top of it. The tenor voice dropped to a deep growling bass, she went round once, leaped off and capered round the bed. Oh, this was a great new trick! And she had just jumped onto it again when I reached her and lifted her onto the floor. But she was now in the mood for games; she boxed a few rounds with Moobli's tail, and even chased a buzzing fly across the bed for the first time.

I was delighted to see her spirits restored but was still worried that the long drive down to London would upset her. But I had no other choice. Her sleeping den box would have to go with us, so would a new litter bowl filled with sawdust and hay – the baby bath was too big for the crowded Land Rover – which I hoped she would use. When we boated down the loch, into the first rain in nearly two months, Liane *maued* and pushed her big claws through the ventilation holes in her box, just as Patra used to.

I did not want to have to leave Liane in the Land Rover amid the noisy fuming traffic of London for the few days of our stay, so I telephoned the owner of the Hampstead hotel I'd once stayed at before and who had accepted Moobli. 'Of course we'd *love* to have a young wildcat here,' he said. 'We have two rather 'wild' cats of our own. Just so long as it doesn't bite the maids – we have a hard time finding good ones!' He was an understanding man and liked animals. But how would Liane take to the long jolting drive, the racket of trucks whizzing past on the motorways? As we set off I watched her carefully.

Far from being ill, or even dying from fright as I'd imagined possible, she thoroughly enjoyed the 570-mile trip. She put her paws up on the shelf by the windscreen to peer out at the passing trees and roadside. She capered about inside the truck, throwing herself in front of Moobli's dignified head as he lay on the bed at the back, and clouting his chin or clipping his ear as she passed. When she wanted to spend a penny or tuppence, she hopped over the seats and squatted elegantly, tail upraised, in the right bowl, even covered it up again. A swift brief uplift of the front hot weather panels cleared the air, and she was no problem at all. Once she even jumped onto my lap and put her paws on the wheel to peer ahead, as if she was driving.

We had set off in the afternoon and reached one of my 'dormitories' by late evening – a gravel pit outside Dent in Yorkshire. I cooked supper on the camp stove, fed the animals, banished Moobli to sprawl over the three front seats overnight, while Liane slept as she often did on the bed beside me outside the covers. As we completed the drive next morning it was entertaining to see the different expressions of car drivers who passed us when they saw a wildcat staring at them out of the window.

Liane took happily to the thick carpeted luxury of the hotel room, just as Moobli had assumed it part of his aristocratic heritage on his first visit. She ate and drank all I gave her and never once made her toilet anywhere but in the litter box. I woke up at six next morning, the sun streaming in the window, and found Liane lying on her back by my side, her paws bent on her chest and her head resting on the pillow. I'd had a few odd girl friends in my time but this was ridiculous. A foot in my eye, a claw in my mouth to wake me up. How charming!

For the next nine days as I dashed about my appointments – with the *Reader's Digest* in Mayfair, two professional photographic studios to have my photos developed and printed, the British Museum of Natural History with my butterfly, moth and insect specimens, and two publishers – Liane and Moobli entertained each other in the Land Rover. As soon as folk heard I had a young wildcat outside, out she had to come and I had to lug her box from the parking meters down long streets and into different buildings to show them. Liane saw the insides of more offices in those few days than I had seen in my first year in London as a cub reporter. And she rose to each occasion, putting on hilarious performances, growling, purring, chasing paper over desk tops, delighting the various office staffs. She really seemed to appreciate an audience to perform for, rather than just plain old me. After my last visit to see a publisher's editor, who showed interest in two of my books, I gave her a lift home on my way to visit friends; I still think Liane, nestling and purring in her lap most of the way, played a fair part in having my first wildcat book commissioned.

One late afternoon, however, we came close to losing Liane for good. I had stopped as usual to throw sticks for Moobli in a green grassy place he liked – the area of Hyde Park by the Serpentine Bridge. As he chased and brought back the small branches, something made me look over the road at the Land Rover. To my horror I saw Liane standing on the driver's seat peering at us, and then squeeze through the slightly open window, pushing it back with a strength I did not think she possessed. The next moment she was down and running across the busy road through the fast-moving traffic towards us. A taxi clapped on its brakes, tyres screeching, but he would not have hit her anyway because she made a huge leap (I measured it later) of fifteen feet into the air, landed on the grass, and terrified by the screeching noise and the great black object that had borne down upon her, scooted straight up a sycamore tree to its first fork – where she stayed.

Moobli saw her too and reared up the tree, whining. One by one, seeing us staring upwards, passers-by came and before long a small crowd had collected. 'What an odd looking cat,' said someone. 'I've never seen a fawn striped cat like that before,' said another. But I played it all down. 'Oh, she's just an ordinary kitten,' I said. 'She'll come down if we move away,' and I left myself. Eventually the gathering dispersed, but it was dusk before Liane finally came down, this time climbing backwards finding footholds on the pieces of rough scaly bark. Then she ran to us with her squirmy *'Chirr'* note. But it had been a close shave for her with that taxi.

On the way home, we visited Edinburgh Zoo Park because I wanted to see Freddy, the tom kit I'd let the Zoo have to be companion for their female in the spring. Unfortunately he had not yet mated with her, probably because he needed some months to settle down. Nor did I see him. 'He comes out at night more, spends most of the day in the den at the back. But he's in fine shape,' I was assured by Nobby Clarke, head of the animal department. (Nobby had been head keeper of small mammals at London Zoo when I had taken delivery of Freddy's father, old Sylvesturr, in November two years earlier.)

Later I was with Nobby in the office of Roger Wheater, the Zoo Park's director, and the two men were discussing a forthcoming TV show to be filmed in the Zoo with Johnny Morris. They were wondering what species he could handle before the cameras, maybe the squirrel monkeys.

'Why not let him handle one of the wildcats?' I suggested facetiously.

'You'd have to leave your tame kitten then,' laughed Roger. 'We don't want anything that's going to rough him up!'

When we reached my home loch next morning and the smaller of my two boats which I'd hauled up into the woods, strong north-east gales were blowing noisily over the hills. But all I wanted to do now was get home to Wildernesse, and certainly Moobli and Liane had had more than enough of living in the Land Rover. I hauled the boat down over branches, screwed on the small 4-horse power outboard, carried eight loads of materials and gear down to the shore and put everything in the boat. Then I set Liane in her box on the centre seat in front of me and, with Moobli in the bow, set off.

At first, the oncoming waves were not too bad but as we reached the bend leading into the wider middle mile and a half of the loch, the waves were deeper, white-crested, and the seas were running heavily against us. I realized I could not fight the waves directly with a laden boat and the low-powered engine, so I slid partly with them to the north, then cut east again where the curve of the land point ahead reduced the power of the waves. Even so, with the engine full out, we only banged along at half walking pace.

Near the steeper rock bluffs on the shore, the waves piled up into deeper troughs and the bow nearly went under each time it slid down one and hit into the next. But we made it to the lee of an island where I re-filled the tank with petrol. We had just headed past Sandy Point and turned north-east for the main three mile run to Wildernesse with the gales now dead against us, when the engine suddenly cut out. I frantically got it going again by using the choke, cursing the fact that I had left my valuable long oars in the cottage, and off we banged again. We had just passed the second curving land spit about two miles from home when the engine faltered and cut out again. This time nothing would make it re-start.

I grabbed a frying pan from one of the bags and paddled desperately to keep the bow to the waves but the boat swerved slowly broadside, was totally swamped by the crashing water, and down we went. I clutched Liane's box by its carrying handle, one of the two briefcases with my most valuable possessions – my wildlife notes, diaries and costly photos – and leaped as far as I could from the sinking boat towards the shore. I was twenty-five yards from the beach, and struggling in about fourteen feet of water. For a few moments I thought I wouldn't make it. Then I remembered. For years, whenever Moobli and I had been swimming together, I had trained him to swim near me, then I'd grab his tail and make him tow me towards the shore, his natural reaction anyway when his tail was seized. 'Come here Moobli. Come here!' I gasped as he paddled around with his powerful web-footed trudgeon stroke. He came, eyes white and snorting like a bear, and just as he swerved off again I grabbed the end of his tail with the hand that held the briefcase. In this fashion, trying to hold Liane's box high with the other hand, kicking madly the hardest breast stroke kicks I'd ever made, we reached the gravelly beach.

Gasping and spitting out water, heart pounding as if it would burst out of my chest, I just lay there for a few minutes regaining my breath. Poor little Liane was crouched in her box, soaked to the skin and terrified. Luckily only a few inches of water had gone through the upper ventilation holes, and I tipped her box up to let it all drain out.

Miraculously the boat had done a nose dive, keening in sideways, bow first towards the shore and its rope was lying snaked in only six feet of water. I couldn't get wetter than I already was so I peeled off my soaked and heavy jacket, dived for the rope then heaved the boat slowly towards the beach. Luckily the heavy items were still all in it and my pack and lighter objects had floated onto the land spit. I unloaded everything off the boat and stacked it all high on the beach, let the water out of the briefcases but couldn't dry the photos because of the gales. The boat, filled with water, was too heavy to haul out or even tip up so I would have to come back with the bailer.

We now had to walk the rough route to Wildernesse with Liane and my essential belongings, then come back with some oil – for I felt the engine had broken down because I had failed to put enough into its fuel mix. I couldn't carry Liane in her big square awkward box for two miles, so I took the top items out of my pack, put her bedraggled and oddly quiescent form onto some dry clothes and strapped down the top. Then with pack on and both hands filled with my heavy briefcases, we set off through the lochside alders and over the treacherous tussocks and rocks.

Halfway home I set the loads down for a rest and to give Liane some air. She had gone! 'Didn't you see her?' I yelled furiously at Moobli who had been walking behind me. 'You damn fool!' She must have jumped straight out into a tree for him neither to have seen nor scented her. I looked back the way we had come and *maued* loudly – but there was no answer. And the landscape was empty.

'Track her Moobli,' I shouted to the shivering dog. 'Track the pussy coots!' He tried but picked up no scent at all. He looked miserable, all in, and I realized I should not have shouted at him.

Angry at myself and shivering in the cold winds as if an electric charge was running through me, I knew there was nothing for it right then but to get my valuable notes home and spread out to dry fast, change into dry clothes and return with oil for the engine and a bailing tin.

After all the long drives and hectic times in London with Liane flourishing when I'd least expected it, now, at the very last moment, she had escaped into the wild just two miles from home. Poor old Moobli. Poor little Pussy Queets. What a damn sad and silly day.

CHAPTER 9

None of this really hurt, and she was such a lovely animal I could forgive her almost anything.

When we returned to the land spit with oil, bailer and a broken oar for a paddle, I'd given up hope of finding Liane because Moobli seemed totally unable to track her. But I searched high and low over the steep wooded sides of the lochside hills *mauing* like a mother wild-cat gone crazy, then to my surprise I eventually heard little answering *maus*.

Liane was up a thick dying alder tree forty yards above the shore. Keeping Moobli back I called softly to her and as I went close she scuffled down to me, clinging hard to the lichen-covered trunk as I lifted her off. It seemed a chance in a million that we had her back, for I felt she was not yet old enough to have found her way back to Wildernesse alone. I put her back in her box, bailed out the boat, and put oil into the engine fuel. But as I was loading up again another mishap occurred. My second briefcase, a moth-eaten relic, slipped from my over-full grasp, burst open on a rock and the gales blew away several sheets of photos. I chased over the beach and hoped I had rounded every one of them up. Then I kicked the boat off as hard as I could, leaving Moobli to run along the shore.

After two hundred yards the engine cut out again. Drifting in the tossing waves, with the boat now light enough to avoid being swamped, I removed, blew out and re-fitted the petrol pipe. Still no good. The gales had blown me back another quarter mile, but I managed to paddle half sideways into the slight lee of a small bay where I unloaded everything again and stacked it all below the shelter of some trees.

I now had to walk back yet again with another heavy load. This time I didn't put Liane into the pack but in a pillow case tied round my neck. She must have been as tired as I was for she relaxed and even went to sleep cradled in one arm, only waking when I had to use that hand for hauling myself up the steeper bits. It took the rest of the day just to dry everything out – my notes, diaries, camera, lenses and sheets of photos, and each of the glass slides had to be individually opened and dried. The final disaster was to find that two sheets of photos, one containing rare pictures of wild otters swimming, the other of Cleo and her young kits together, were still missing. I never did find them.

By nightfall Liane was in fine spirits again, eating and drinking heartily, playing in the study with all her old toys, the end of the bed-spread, the letter files and scuffing out the hay in her plastic baby bath.

Next day I carried the heavy 20 horse power outboard down to the shore, hauled the big boat down to the water and went back to retrieve everything and the small boat. It was a sunny morning so I had draped my soaked clothes and laundry over the bramble bushes that were now filled with blackberries. Even my breakfast cereals had to be dried out in the sun. It was now mid-September and the single remaining branch of the old plum tree I had heavily pruned and manured during the winter bore such a bountiful crop of ripening fruit that it was bent low to the ground. I picked seven pounds but, fearing it might still break, propped it up with a long pole.

On September 16th Liane caught her first shrew in the bracken jungles fringing the east wood. She did not play with it half-alive as a domestic cat will do but brought it into the study dead. She did play with it then, however, throwing it high into the air with both paws and batting out as it fell again. To my surprise, for few animals will actually eat the musky scented shrew, she eventually ate most of it on the carpet, leaving the snout, flanks, tail and part of the rear quarters. She was now three months and ten days old. Due to the artificial environment in which I had kept her through kittenhood, plus the London trip, she was slightly more backward at hunting than she would have been in the wild.

My plan now was to allow her full run of the house and the outside too, so she could choose to come in when she wished yet also hone her hunting skills naturally. It was clear that her system needed live prey. This plan was hastened along when I woke up next morning. My olfactory senses indicated that while she still used her litter box in the study-bedroom, her increasing size and weight (four-and-a-half pounds) plus her high protein meat diet were producing toilet performances of such magnitude and pungent odour that a heavy strain was being placed on our close indoor friendship!

This is not a subject for lengthy discussion, though it is as much a problem with animals as it is with young children. But during the next five weeks my efforts to 'house train' Liane completely ranged from temporary success, hilarious incident to total frustration. She regarded the study-bedroom as her true home – the great outdoors was only for hunting but her loo was indoors! I now had to try and reverse the habit.

At first I put her litter box in the kitchen and while she was shut in there with Moobli she used it. But when she was outside during the day she would suddenly hare into the study and, finding her litter box not there, jump into the cold ashes of the fire!

On September 18th I watched from the windows as she stalked a short-tailed vole at the rear of the cottage. She stayed still while it came out of its hole and wobble-tottered along as if on tiptoe, tensed her muscles then sprang with both clawed paws stretched out, trapped it down, pulled it back for the death bite on the skull and dropped it. All the while her pupils were wide and as black as jet despite the daylight. I wanted to see if she would let me have the vole but she flared, growled, picked it up in her mouth and still growling ran into the woodshed where she ate all but the tail and part of the little rear haunches. I didn't mind for it was good to see her skills and new independence increasing.

But the toilet training continued to present problems. For a few days I shut her overnight in the kitchen with Moobli, leaving the back door open so both could go outside when they wished. But now she performed in the shavings on the workshop floor – so I swept them all up. Then, with obviously no protest from him, she used Moobli's bed! I decided to try more forceful means. I let her sleep as of old in my room but when she went to perform in the fireplace, I tapped her behind lightly as I could in play when she became too boisterous, and immediately put her out. This treatment, of course, failed utterly, for it wasn't play and it just made her wilder. So then, as soon as she showed any toilet signs, I hurried her into the pens and shut the gate, feeling she would *have* to go outside there and would maybe learn that outdoors was better for such purposes, and that indoors was just for sleeping, playing and where she could expect some caresses from her foster parent – for she still leaped on me for these every evening. One of her favourite games now was to lie in the valley between my knees, completely upside down, her head hanging over the edge, belly exposed, legs in the air and staying totally still as if she were dead, while I tickled and poked her gently. She was not in the least bit ticklish.

She was now catching large insects like bluebottle flies which she chased with lightning dashes, her claws slashing in round house style until she struck them down, then she ate them. Insects appear to make up a fair part of the diet of young wildcats, and the speed and reflexes needed for a successful catch stand them in good stead for serious hunting. Often she tried to capture the great black and blue Aeshna dragonflies, as they soared and darted almost contemptuously a yard or two beyond her reach, for with the ten thousand lenses of their extraordinary eyes, fast reflexes and flight powers they could easily see her coming and dodge away.

One afternoon when I was sunbathing by some logs I'd cut and left to mature, I suddenly felt sharp pains in my left hand. Liane was under the logs and making mock attacks on my fingers. She was unusually playful, throwing herself on her back and swiping out with paws at full stretch, rolling over and over as she bit a finger lightly. None of this really hurt and she was now such a lovely animal I could forgive her almost anything. When she had finished playing she climbed right onto me, her foot pads cold on my skin, and lay with her warm furry head against my neck. Such blissful and trusting moments were more than reward enough for the problems and long hours of patience.

On September 26th, after I had seen her catch two more voles, Liane went voluntarily into the pens. Again I shut her in but this time for the night, hoping as she had now *chosen* to enter she might regard *them* as a good toilet area. It did not work, for although she *maued* raucously and leaped into my arms next morning, when I went to cook lunch I found she had used her hall litter box for the main deposit but had made a large puddle on my fire lighting papers. She spent the afternoon climbing among the east wood hazel bushes, working up a huge appetite. After eating, she finished the usual hand caressing session in the study by furiously licking her forelegs and shoulders then went to try and soak my letter files. She squawked indignantly when I picked her up and put her outside none too gently.

On the 28th we invented a new game. I picked her up and threw her towards the thick larch pole of my wind generator – she sailed through the air, landed and clung on as if it was a magnet. Then she climbed down – and ran back to me to throw her again. A few more tricks like that and she would be a music hall turn! She had also developed a propensity for giving me sudden surprises when I wasn't looking – such as leaping onto me out of nowhere when I was leaning against the porch logs for a little sun on my face after hours of typing.

I now tried new toilet training ploys. I fed her every meal in the pens so she began to go into them when hungry, but when she sneaked

By late September Liane was hunting regularly, but she often came mauing through the window for a cuddling session before leaving again.

back into the house to 'perform' I held her close to the mess, shouted 'Bad girl!' and clapped my hands, sending her high-tailing outside again. This worked – for about two days. Then I found her one morning making a large pool on Moobli's towel on the trestle seat I'd made for the kitchen table. Performing on clothes and towels was really too much so in the hope of discouraging her from at least doing that, I pulled the towel from beneath her. She squawked and shot outside. That night she did not return for food and was not in the pens or the woodshed. During the day a huge golden eagle had flown low from east to west behind the cottage, pursued by two ravens who *'krok kroked'* above and darted down but not too near it. The eagle had taken no notice of them whatever and had just kept soaring and beating lazily along. Scared the eagle might just come back and spy Liane or that a big fox would get her, I set Moobli to track her. We eventually found her in total darkness by torchlight – a good forty feet up in a thick oak tree on the north hill. She *maued* in return but refused to come down.

In the morning I woke to hear loud *maus*; I got up to let her into the house – she ignored me and her litter box in the open hall and promptly ran to squat in the letter files. I pulled her out by the base of the tail, was bitten on the hand, and from the size of the pool she made outside she must have been saving it up all night. Moobli, too, was feeling the way I did about all this and that afternoon he came whining into the study. I went out with him – to find Liane at her toilet on his bed again. As she fled outside at my entrance I was glad that he at least had not got angry with her. I was beginning to learn that she just would not seek out a litter box unless she and the box were kept in the same room all the time, but I decided to persist a few weeks longer. That night I set litter boxes in the hall and every room.

It was now early October and the red deer stags were roaring on the mountain slopes as they trotted round the perimeters of their harems of hinds. On a trek, Moobli and I paused behind a rock to watch a nine-pointer chase one hind in a half circle, calling short barking *'but but buts'* as if spitting at her, then he turned off to one side, stretched his head and neck out straight and roared his challenge across the glens. The large, brown and black banded northern eggar moth caterpillars were thick on the ling and tussock grasses for the last feeding before spending their second winter as pupae, and in boggy areas the tips of soft rush were turning yellow. The alder leaves had been falling from the trees along the shore for three days now and the first beech and ash leaves began to drift down too. The next strong winds would start the spectacular falls of autumn.

When we returned Liane came into the house briefly as if just saying hello then went straight out again, and with delight I saw her at her toilet in the pens. When I approached with food she dashed to the front fencing with a loud *'Brroo'* chirr, clawed playfully at my hands, biting the soft part between thumb and forefinger. She looked healthier now after her long spells outside, her fur had grown thicker and her eyes were almost as large and black as old Sylvesturr's.

Two nights later the first strong south-west gales began, sending five-foot high waves crashing down on the shore. Although Liane seemed quite happy in the open pens, I felt the winds too strong to leave her out permanently until I'd overhauled and modified the den and made it more draught proof. I brought her into the kitchen overnight and this time she used her litter box. Perhaps she was learning at last. When the gales died down next day she went out to range both woods in spite of slight rain but she caught nothing. The mice and voles were probably keeping low after the loud storms and saturated ground.

On October 9th, in the light of a weak watery sun, Liane showed me a new dance routine in the open area between the cottage and the pens – the first time she had danced out of doors. With ears back, toes pointed outwards like a ballerina, she hopped, ran and drifted about, pouncing on leaves and tufts of grass, her tail high but curved and fluffed out to almost three inches thick. At times she looked like a mad earless hare. I shot off a whole roll of film, standing well back so that she felt quite free. I had to use a 300 mm lens and it was hard to focus on her flying form. When I bowled a small apple towards her she chased it like a dog, cupped it in her hands, bit into it, then leaped away and stalked it again, stiff-legged, pounced once more but did not bite then ignored it thereafter. The tart acid taste of the fruit obviously did not suit her palate.

At night she came in through the study window for a short play session but now she was stronger her claws really hurt and my hands

In weak, watery sun, after the first real autumn gales, Liane danced out of doors for the first time. She hopped, ran and drifted about, tail fluffed out like a mad short-eared hare.

were soon bleeding. When I stopped moving them, as if they were 'dead', she let go and pranced away. Later she sat by my typewriter on the desk watching the moving carriage, then lay on my lap, both fascinated and puzzled by my pipe and watching the smoke come out of my mouth and curl upwards into the air.

All next day Moobli and I were away on a stag trek, and, when we returned, Liane's evening play session had an element of real ferocity about it. She glared at my hand, then growled as she charged it, a real karate attack. She clutched, bit and kicked with her rear feet, rolled on her back and flared, certainly not *trying* to hurt but looking terribly fierce all the time. Then suddenly she lay on her side, both forelegs stuck out stiffly, and let me stroke her chin.

On the 12th of October her bad toilet habits began again. She ate her evening meal, climbed into some new blankets in her bed box, raised her tail and let go. Two days later, after I had just returned from watching the first of the season's redwings – recently arrived from Europe to escape the continental winter, winging in to raid my rowan trees for their scarlet berries – I found Liane had mis-behaved again. This time on *my* bed, and she had ignored all the litter boxes! Next afternoon when I had been distracted by the roaring of a stag and had gone down to the beach to get a clear view of it on a low hill to the west, Liane, who had been outdoors all day, sneaked into the kitchen and puddled right in the centre of the carpet. I could not just sit back and let her get away with it, and, even if it meant her becoming wilder and more alienated, she would have to learn. So I pushed her nose in it, was rewarded with a loud spit and claws in my arm, and with the gauntlets on I shoved her outside again. During the next few days she kept it up, and so did I. But it seemed then that she was starting to accept her nose being pushed into it as a sort of ritual, the price she had to pay for the pleasure of using the cottage as her loo. It had no deterrent effect whatever. On stormy nights she ate in the kitchen, looked fierce and wild when nature's urge overtook her, looked for somewhere out of sight to go, went, and only *then* ran out of the house.

For the next three nights Liane ranged along the area of the shore and now the water and waves no longer scared her. She was evidently catching plenty of food for she ate barely half the meals I gave her. One night we went out to track her, to find out where she went, and Moobli treed her up a far larch in the east wood, a full three hundred yards from home. She slept by day now, usually in the den of the open pens where she had been born, hunted most of the night, only coming back for a meal if she had been unsuccessful in the early morning. It was strange that she never used the woodshed as a headquarters as had Cleo and Patra.

One morning she came in and, as I was preparing her food, ignored her boxes and used my best clean huge beach towel for her major business. I controlled myself, washed the towel with silent curses as she dashed outside again, then fed her in the pens. I'd just about decided to give up when one afternoon after lunch I'd gone into the kitchen and, like the proverbial fat man who doesn't see the banana skin, put my foot in one of Liane's brand new giant offerings and nearly brained myself on the concrete. As it was I merely pulled a thigh tendon. She was still in the room and my fall scared her so that she leaped to the window seat spitting and flaring.

Two days later I made the final try. I felt that perhaps she would not use the hay of her litter boxes now as it was the material of her bedding in the pens, so I put down one box with sawdust in it, one with bare earth and another with grassy turf. Then I brought her into the study to see if we could recapture the careless joys of the days we had shared in the room when she was a small kitten. I fed her but within two minutes she had climbed right onto the bed, ignoring all boxes, and was squatting. To me it seemed she was keeping *her* bed clean yet was prepared, almost as if deliberately, to make a mess on what she *knew* was mine. Somehow I kept to my new resolution not to ever try to discipline her again.

Liane knew she had done wrong. She looked at me with glowering eyes, hissed as I got up from my chair, then confused between me as authority, me as foster parent and me as threat, ran about trying to hide. She now looked just like old Sylvesturr, the same ferocious gloomy expression on her face, the jutting lower jaw so characteristic of the wildcat. But I just kept talking to her soothingly, wheedling my way back into her confidence, and in a quarter hour of stroking and fussing, she was finally purring again. I then quietly let her out of the door and cleaned up the mess.

I accepted then that the bedroom days were too imprinted, the earlier training to the house just too successful. It was impossible to reverse it all now so that she performed her toilets only outside and never in the house. As a kitten she had used her litter box regularly when it and she had been at all times in the one room but, once she started to range outside alone, the choice of rooms, inviting surfaces and boxes when she returned indoors, was just too much for her.

My fondly nurtured image of her returning from hunts for only caresses, play sessions and a snooze on my lap went up in smoke. I had to admit defeat at the last hurdle – although I had hand-tamed her, I had totally failed to house train a wildcat.

CHAPTER 10

Winter can be hard in the Highlands, but, at six months and now weighing over five pounds, Liane was capable of fending for herself, though I still fed her every evening in the woodshed.

From now on Liane's visits indoors would need to be supervised and her main headquarters would have to be outside in the pens, as she seemed to prefer them to the woodshed. As it was nearly the end of October, with winter only weeks away, I decided to improve the rocky den for her. First I drove a square post into the ground and nailed a new aluminium roofing sheet lengthways over the top of the den so it projected three feet at the front as a protection from rain. Then I built a second door about a foot inside the den with its hole set on the opposite side to the entrance on the main door – so whichever way the wind blew there could be no draughts. Loads of bracken were laid two feet thick over the roofing sheet and I filled the den with new dry hay, stuffing it into every tiny crevice so it was really warm and cosy. Liane took to it immediately.

Late at night, however, as rain began to fall, she came *mauing* loudly round the porch to come and sleep indoors as she usually did on wet windy nights. This presented problems. I didn't want to banish her completely from the cottage and our company when she herself sought them. Nor did I want to shut her totally in the pens in case she became bitter, morose and fierce like old Sylvesturr. So I compromised. For the first few nights I put Moobli's bed blankets in the little hall, shut the kitchen, rear room and study doors but left the front door open.

When I went out early on the next two mornings Moobli was sprawled out asleep and there, lying right across him, was little pussy poos herself! Now she was denied my proximity as a sleeping partner, she had decided to adopt Moobli as a replacement and he, far from minding, seemed inordinately proud of her choice. I fed her in the pens, feeling this would be the normal pattern from then on. She still liked the pens as a daytime refuge for she made a nest in the bracken on top of the aluminium sheet and on sunny days dozed for hours, basking in the warmth, safe from any sudden disturbance from Moobli.

It was odd that she had chosen to sleep across him in the hall for when they were outside her attitude to him was much the same as Cleo and Patra's had been when he suddenly loomed nearby – she was scared. Once when I had not seen her all day, I set Moobli to track her. He found her in the large rhododendron bush by the path and when she ran to some old ruined walls near the west wood, he naturally followed. She was bigger and heavier now and, it seemed, could keep him at bay, for when I caught up I saw his tongue was bleeding in front. But as usual he was showing no belligerence, just wanted to get close and sniff her, and she was letting him know that while friendship between a huge male Alsatian and a young lady wildcat was possible, it had to be when *she* chose and on her terms.

One afternoon I gave her the fresh carcass of an ageing red squirrel which, sadly, had died in the woods, to see if she regarded it as food. She came down, sniffed it all over, pushed it briefly with her paws, then carried it up to the den roof bracken, pulled it under her rear end and went to sleep. Later I saw that she had taken it to the ground and buried it under dead bracken. She left it alone all next day, and when I went out with her food around 7 pm I found she had disappeared. Tracking her with Moobli revealed that she was hunting in the west wood for the first time.

Evidently she caught nothing for next morning she had eaten part of the gut of the squirrel. When I went out at 3 pm she had eaten all the gut, the lungs and much of the back meat. As I watched she then crunched through the spine and ate most of it, leaving eventually just the head, forelegs, rear legs and tail. Her preferences seemed quite different from humans, who tend to eat the leg meat of an animal first. She ate the heart last, then covered up the remains – as wildcats do for protection from crows, ravens and insect and slug scavengers.

When on November 2nd the first winter snows fell upon the high tops of the mountains from the 1,500-foot level upwards, and drizzle set in for three days on cold north-west winds, Liane made a voluntary change in her sleeping quarters and started using the old hay box in the woodshed for the first time. She came indoors occasionally at dusk, after *mauing* at the study window, for brief stroking and purring sessions but as soon as she was set down on the floor again she showed no desire to sleep on the bed or the sea chest sweaters, and leaped onto the window shelf to be allowed out. She still used the pens in the daytime and some days, provided I told Moobli to sit, I could actually call her out and she ran to be picked up and fondled with her loud chirring *'Brroo'* of greeting. But I left her alone for long periods now as she seemed to be adjusting well to her new life outside, and in her new found independence I felt if I bothered her too much I might just lose her.

By November 15th, the nights were drawing in early so that I had to light the paraffin lamp by 4 pm and, with south-westerly gales blowing driving rain into the pens, Liane transferred to the woodshed as her main headquarters. Yet, despite her toilet problems in the cottage, she kept the shed floor clean. I was delighted by this for now, when I went away on a brief research trip, I could lock her in the shed with a food and drink supply, leaving special exit holes for her to go in and out at will. That way she would always have a safe refuge – which she would not have had with the gate tied open in the pens. So from then on I fed her, just one meal in the evenings now, only in the woodshed. For the rest of that winter, Liane ran totally free. But I still had many chances to watch her antics and interesting behaviour.

One sunny afternoon I saw her staring intently at the ground in the pens. A late dung fly had been caught by a wolf spider that had darted out from the shelter of a leaf and leaped upon it. The spider never got its meal for Liane seized them both in her mouth and chewed them up. She also took to leaping on the window sill at night to catch the anomalous moths and the sturdy brown December moths which were common in my woods and had been attracted to my light. She knocked them down with her paws, then chomped them up like breakfast cereal.

During the first few days of December winter laid its bitter chill upon the land, sleety snow fell heavily, straight down, dark and grey like flakes of zinc. The clouds, blue-dark over the iceberg mountains, allowed little light on the snow that now covered everything down to the 700-foot level and lay in patches over the ground outside Wildernesse. Webs of white mist hung wraith-like over the glens and my water system froze up, forcing me down to the loch for buckets of water. But for the wildlife watcher, snow has one wonderful quality – it preserves animal tracks, often for days, until it melts. Finding Liane's tracks outside the woodshed, we followed them through the upper slopes of the east wood, past the water pipe in the pool below the first waterfall and up to a small mossy platform on the bank of the pool above. There some scattered feathers told the story. She had caught what looked like a thrush.

Back in the woodshed she lay asleep in her hay box, murmuring a lazy chirr as she stretched out in her box. But on the floor lay not a thrush but the remains of a migrant redwing. Oddly she had eaten only the wings, tail and head, leaving the body, which was pitifully thin, the breast bone sticking out as sharp as the back of a knife. Nearly all the flock of a hundred and thirty redwings had stayed in my area for only a few autumn days, then when the rowan berry supply had been exhausted, they had migrated for the richer feedings on the farmlands to the west. But I had noticed that four had stayed behind, their wings held low as if too weak to fly further, overturning leaves in the woods in their search for worms, insects and grubs. And when I had walked nearby they had merely fluttered weakly into the nearest bushes.

Liane had caught one of these debilitated birds, probably destined to die anyway, who had shown no wish to join the droves of tits, finches, robins and blackbirds now flocking to the bird table. Although redwings are common migrants to Britain from northern Europe in winter, they only began to nest in small areas in Scotland in 1925, and even today there are probably no more than fifty breeding pairs there out of the whole United Kingdom. I was glad it had not happened in the spring.

The cold made Liane hungrier and she was gaining weight too, just over five pounds now, with hefty shoulders and forelegs. When I put her into the pens during a sunny afternoon spell, she sprang and tore holes in my hands with her rear claws. Despite this brief sunshine, icicles persisted in the burn, hanging in clustered pendants from the rocks, and my garden cabbages were as hard as cannon balls. As they were now deep frozen naturally, and would rot if a thaw came, I put a dozen into my calor gas 'fridge.

Next day Liane did something she had never done before. She sprang onto the bird table after a great tit, which escaped easily and denounced her with scolding, scissoring notes from the safety of a little willow bush. I snapped a towel at her and she snarled and dashed into the pens where I shut her in all day in the hope she would get the idea the bird table itself was out of bounds. Possibly she did for she did not leap on it again for well over a year.

In the early hours of December 10th, with heavy snow falling all night, I was woken up by a loud penetrating barking scream *'Kaah Kaahf'*, halfway between a bark and a cat being strangled. Despite the bitter cold I dashed to the window but could see nothing in the near-dark. Next morning Moobli and I tracked all over the front pasture and the two woods. We found big padded five-toed badger tracks leading from the west wood to the east – where we lost them because there was little snow under the trees. Here and there were small scrapes where the badger had scratched for worms and grubs under the soil. Then we made a more surprising discovery – large wildcat tracks in the snow, measuring just under one and a half inches across and in places nearly four inches deep.

They could only be Sylvesturr's! The tracks started from the east wood and went to the rhododendron bush south-east of the cottage where they vanished into the snowless areas below the thick broad leaves. But Moobli scented around, stopping when I told him to as I did not want him stepping into the tracks, and we picked them up again on the edge of the glade below the huge Norway spruce that dominates the east wood. They then went to another rhododendron bush to the south, headed across the pasture, crossed the badger tracks below my vegetable garden but then vanished among thick swathes of brown bracken. Most astonishing of all was that halfway across the pasture, Liane's smaller tracks, often barely breaking the snow surface, had come from the east edge of the cottage to join his and went side by side with them to the garden. Here she had separated and headed north again, back past the cottage and to the woodshed – where we found her asleep. Had she actually met Sylvesturr or just gone the same way at a different time? Could the old tom have come back looking for Cleo, or was he now interested in Liane? Whatever was happening I was delighted by this evidence that the old curmudgeon was still alive.

That evening I broke my resolution never to try and catch him again and set the box-cage trap near the lower rhododendron bush. At least I could check his condition and if he needed it put him in the pens and feed him for a few days before releasing him again. I doubted the chances of success, for he knew the trap well and had a memory like a computer. At least we could track him again next day. Liane was now extremely active, darting about and playing in the snow while I tried to take pictures. She went to the toilet and I was surprised to see her carefully cover the droppings up, raking a mound of snow over them with both front feet.

Next morning we found Liane had been in the trap, which had not worked, her smaller tracks leading to and from it – all the meat had gone. She had clearly scented the meat from over a hundred yards away because for four snowy days she had certainly not been in that area. But as soon as I had put the meat in, with the slight breeze now coming from the south-east, she had found it. After my experiments with the other wildcats, it now seemed conclusive that these animals do use scent for finding carrion in winter. But there were no further tracks of Sylvesturr's in the snow patches that were left after several hours of a fine, sleety drizzle.

Overnight on December 16th Liane killed a blackbird in the spruce glade and had taken it to the woodshed to eat. The blackbird population had built up to an unprecedented eleven around the cottage that winter but, where these birds are usually fairly tame in urban gardens, these were wild and shy. They did not just fly to the bird table and eat quietly like the other birds, but each advanced separately with loud nervous *'cup cup'* notes, waking me in the twilight before dawn, ready to fly off with frenzied shrieks into the bushes at the slightest movement. The one she had caught was a rather silly young cock with a bedraggled tail and poor plumage, and as with the redwing she was merely following nature's law where the weak, foolish and least wary are taken first. I was still worried about the remaining redwings and shut her in the pens overnight. As if she knew my intention, Liane clawed at the fencing and looked really sulky as I wired up the gate.

She did not like such incarceration any more than had Sylvesturr, and in the morning she ran to me with loud *maus* as I undid the gate again, and actually leaped into my arms. But she spent the rest of the day away from the area and was not back at feeding time. When I went round the west wood with Moobli he picked up her scent but she must have hidden herself well, for we were well past her when I saw her haring up the path to the woodshed. By shutting her in the pens when *she* did not want to be in them would only alienate her further. No, I decided, she would have to stay free, and the birds would have to take their chances as in the wild. But I was glad I had not kept all the wildcats in the woods.

Two days later, finding she was not in any of her usual haunts, we hurried round the woods to find her – without any luck. When we returned, however, I found she had gone into the cottage of her own accord and was sleeping on my bed. Suspiciously, I sniffed the air – nothing. She lay looking up at me with half closed eyes as if saying, 'I can come in and behave you know!'

Christmas Day was bright and sunny and rather than sit moping about being alone at the festive season, my few cards – mostly from lonely strangers who had read some of my articles about the wilderness – staring blankly from the log mantlepiece, I took Moobli on a short hard trek up the steep ridges and rock faces to the west. We checked a rocky den at four hundred feet which Mia had used shortly after she had left but there was no sign of recent activity. We did, however, find the new sett of the badger that foraged the woods at night. The loose black droppings in its deep toilet scrape outside the sett showed it had stomach trouble so I left an unwrapped meat sausage on the ground well below, which it could scent and eat if it wished. (This meat on its own had always performed a good stomach binding job for Moobli!) And after taking some pictures, we came home.

That night I brought Liane into the study and the three of us had a fine little party, Liane stuffing herself on turkey giblets, then capering about the floor, more ponderously than she had as a kitten, and clouting Moobli's muzzle with her paws. The only trouble being that as she was much bigger now, she hit harder, and I could actually hear the blows land as he flicked his huge head this way and that.

He took it well for a while then finally decided to enter the festive mood himself and respond in kind. Instead of trying to bite, however, he sat back on his haunches and swiped out with one of his own vast paws, downwards, so that one of her mock attacks ended with her spinning – and flaring – across the room. That was enough of stepping out of her weight class, she must have thought, for she leaped onto my lap as I typed my daily diary and after a few minutes *maued* to go out through the window and into the more real and exciting world of the night.

CHAPTER 11

By early January, Liane had her first good thick winter coat, and she came inside for short play sessions most evenings.

Moobli and I saw the New Year in by camping out in the woods – a yearly ritual of renewing close contact with nature which seemed more appropriate to my wilderness life than drinking heavily and singing Auld Lang Syne to oneself. It had been a bitterly cold night – even the condensation inside the tent had frozen into icicles – so on our return we used a brief sunny period to storm up to the high peaks above to restore some heat into chilled bones. I also wanted to take a picture of Wildernesse in the clear winter light.

As we rounded a small knoll, we heard two huffing warning barks from a red deer hind, and saw a small group of hinds, yearlings and calves running away downhill. And galloping along with them was a fine eight-pointer stag. I recalled that about the same time last year we had seen the stag with the hinds – when it had been a six-pointer. It was an unusual sight, for after the autumn rutting season adult stags leave the hinds and go off to spend the winter, spring and summer in bachelor herds on different grounds. But this stag seemed to be with the hinds permanently. Wanting to record the event, I wasted valuable seconds changing lenses, though just managed one shot before they were too far away. As we reached the peak for the picture we had really climbed for, a great cloud obscured the sun and plunged the view into gloom.

That night Liane behaved as if we'd been away a month. She reared up at the window, *mauing* and stretching her white jaw upwards until I let her in. Then she lay in my arms as I read a book at my desk, reaching out jerkily to touch Moobli's box-like muzzle with her paws as if to reassure herself he was back again too. She was extremely affectionate that evening and once when I held her up in front of me she placed both paws on each of my closed eyes. She could have taken both eyes out with just one clutch but she just held them there. She did not really like being held up and away like that but she now knew I loved her – no need for violence! Then she stayed totally still, her big golden eyes gazing into mine only inches away. It is quite an experience, gazing close into the eyes of a reputedly ferocious and untameable animal like a wildcat.

I found, too, that she liked to have the nape of her neck and the short hairs of her head given quick hard little pinches. She squirmed with pleasure, giving faint little *maus*, but I stopped when I suddenly realized *why* she liked this – it gave her the kind of feeling she would get from the love bites of a tom during mating!

On January 14th I knew the fine, but bitingly cold spell which had resulted in the water system being frozen for several weeks was coming to an end. The deer, who had been using the west wood for nightly shelter from the freezing north-east winds, moved up the slopes and disappeared over the ridges to the north-east. Deer know better than any man when the weather is going to change and sure enough, the wind switched to the south, drizzle replaced the snow and began to melt it, and the water started running freely again in my pipe from the burn.

Three nights later snow again began to fall, and we used it to track Liane's nightly rangings – she was now prowling along the lochshore before dawn and had established a definite toilet area just above the waves, seemingly to mark one edge of her territory. She still came in through the window most nights to re-establish contact but after half an hour she began staring out into total darkness, her head moving up and down as if actually sighting mice or voles, and *mauing* to be let out again.

One night as we returned home late from a shopping trip by the light of a half moon and I was envying Moobli's warm running exercise along the shore while I suffered near frostbite in the boat, Liane danced right down the path to greet us. But the moment the boat touched the grassy bank, back she hared to the woodshed. Moobli arrived at the same moment and, starting off six yards behind her, had lost a further four by the time Liane reached the top of the path. Far from being scared of the great 'wolf' behind her, I was just in time to see her caper out from the gap between the shed and the cottage wall and give him a couple of light cuffs round the muzzle, as if saying, 'You're just not fast enough, old man!'

It surprised me that fast as Moobli was she could apparently outsprint him over a short run. In the months since we have turned this into a kind of occasional game – I carry Liane down to the shore, tell Moobli to sit, off she scoots, literally as fast as a hare, and off goes Moobli after her. He has never caught up with her or made her swerve into the rhododendron bush halfway up the path yet. Like most felines, except possibly the lion, wildcats have small lungs, and on a long run out in the open he would have worn her down within two hundred yards. But he would never have hurt her – apart from myself she was his only other companion and his feelings for her seemed as strong as mine. On the other hand, if Liane had ever felt herself at bay out in the open, her over-all tolerance of him, unlike the genuine affection a dog or human can feel, would have been instantly replaced by her wildcat instincts. Once, in the spring, I saw what happened when he caught her unexpectedly out in the open front pasture amid the new growing grasses. She immediately launched into the wildcat's characteristic frontal display, rearing high up on her front legs, spitting loudly and smashing her paw down on the ground – enough to set him back a pace or two while she dashed for the nearest bush.

In mid-February, with gales and incessant rain blasting from the east, Liane seemed to have evacuated the woodshed, was ignoring the den in the pens and had hollowed out a new home for herself in the thick bracken below the aluminium sheet at the back of the pens. She had also made a hole on its north side so she could see out. Perhaps she had hoped this would be her 'secret' den for when I peered into this hole her large golden eyes glowered at me with a most unfriendly look. Now the nights were becoming shorter – I could now stick it out until 6 pm before lighting the lamp – I wondered if her move could be something to do with the normal wildcat mating season, and Liane just wanted to be more alone. The trouble was she occasionally stayed in or hunted from it leaving her nightly food untouched in the woodshed. Instead of filling her new den with tightly packed bracken, I decided on a different strategy. I removed the inner door of the main den so she could see out when inside, and also knocked out a hole in the centre of the woodshed wall nearest the cottage.

It was vital that Liane kept using the woodshed as her main headquarters as I would have to leave shortly for a few days. I was due to meet Nature Conservancy chief warden, Richard Balharry, concerning my golden eagle work that year, and to deliver my first book on wildcats to my publisher and to complete some biological research at the British Museum in London for the book's appendix.

Liane was too large and unpredictable to inflict on any London hotel. Besides, now that she voluntarily spent long periods alone, I felt she would be happier with a proper food supply on the territory she had come to know well than in the traffic-filled city.

The new hole I made in the woodshed wall meant she could sneak in along the foot wide gap between the two buildings rather than have to expose herself in the open as she squeezed through holes in the outer walls. I also 'hid' her haybox between the stacked firewood logs. It took her two days to find the new hole but then she began to use it regularly and she also liked the new position of the hay box, as from then on she only used her bracken den for an occasional daytime snooze.

On the late February day that Moobli and I left for London, I arranged six pounds of cooked and raw hearts, sheep flank and liver over the log piles, five 15-ounce sterilized meat sausages on the floor, plus trays of dog and cat biscuits, six pints of long-life milk-and-water mix, a bowl of pure water, and a final touch – a pound of half cooked haddock. Liane had enough food for a month. I said a fond *au revoir*, locked the doors and set off up the loch with Moobli.

When we returned ten days later Liane was fine, though a trifle thinner. She had eaten all the meats and fish, but she had plenty of biscuits and three of the meat sausages left. That night she behaved like a lost child recovering its parents. She came through the study window, clawed up my sweater, chirring and *mauing* and looking up into my face as if to assure herself it was really me after her longest time alone. I tried to put her onto the sea chest sweaters while I typed up all my diaries but she insisted on staying in my lap even as my fingers flailed away.

Next morning she woke me up before dawn, *mauing* so loudly on the window sill I had to let her in again. I had no sooner climbed back into bed, trying to emulate her ability to fall quickly into deep sleep on the covers beside me, than two missel thrushes began to rival each other in song from the topmost twigs of the highest trees in each of the woods. With my head still full of town life I found further sleep impossible.

When I went into the woodshed later that day I was perplexed to find Liane had torn open a large sack of sheep nuts that I kept each winter to help feed any sickly sheep or deer we found. The brown nuts looked as if she had deliberately scraped them right across the floor. Certainly she had not done it to make a litter area for there were none of her droppings in the shed. What on earth had made her do it? Was *she* eating them? A day or two later I had my answer. I peeped over the lower door to find her sitting hunched between two logs, her eyes black and staring, her whole attitude one of keyed-up expectancy, clearly waiting for mice or voles to come and feed on the sheep nuts. There could be no other reason for her ripping open and burrowing into the treble-thick paper sack except to deliberately spread them over the floor in a strongly smelling area so her normal prey would be attracted. While I could have attributed such foresight and cunning to a fox, it seemed extraordinary that such qualities were also possessed by a wildcat. Indeed, by dusk she had two large woodmice lying beside the nuts.

Later I watched how she ate them. First she grabbed the tail and rear feet to pull it closer, then bit down into the skull with her side molars, cracking and chewing it in several places so it was all soft. Then she turned the mouse round and chewed so it slid in directly from the front of her mouth. Now and again she switched it sideways to crunch the spine in her side teeth, but she ate everything, tail and all. It seemed only when she wasn't really hungry that she scraped the gut contents aside or left a few parts.

Spring came late that year after the long hard winter, but when I went out on March 17th I found the first three daffodils were blooming and I saw the large female buzzard soaring over the cottage and south-westwards over the loch. Presently I heard the short high piercing *'Keeyoo'* of the smaller male as it hastened along in her wake, calling loudly and circling round her as they disappeared from sight.

Despite her enthusiastic welcome on our return, Liane became more wary after a few days and far more active than usual, often returning to the woodshed just to eat then immediately taking off again into the woods. Her window visits at night were more occasional and instead of coming in, she just 'fought' my hand through the gap, gave my fingers a few playful bites then dived off again and away into the dark. Several times, at dusk, I saw her in the open areas making odd brief squats on the ground. When, on March 26th, she actually came through the window for a brief cuddling at the desk, I felt I had found the solution to this behaviour. When I stroked her tail area, she exuded a strongly scented sticky liquid that was certainly not urine. She also kicked her rear legs backwards, each one separately, which she had never done before. And a quick pinching movement on the nape or short hairs on her head, which she had always liked, now produced a tail-raised squirmy movement of her rear as if she was ready for mating.

Male wildcats have special scent glands from which they produce a secretion which, mixed with their urine, they spray onto trees, bushes or grass tufts. This is used not only to warn other males to stay away but also to inform females of their presence. It now seemed highly likely that females have a similar ability, hence Liane's short squattings. She was obviously in oestrus and was instinctively leaving her scent around to attract a tom or let one know she was there. Later in the dark that night I went out to find her perched in the willow bush near the cottage, as if waiting for a certain little fellow to pass by! I had found no fresh evidence of Sylvesturr being still in the area, and as wildcat males are monogamous, if he was it would be because he was looking for Cleo. On the other hand, if Liane became pregnant it was heavy odds against any other wildcat being the father as all the male kits had been released miles away.

By early April, the larch trees were sprouting their first light green needles, and in the woodland clearings a few wood anemones shivered their white flowers on weak stems as if scorning the protection of little rocky clefts where primroses emerged to crouch from the wintry north-west showers. The rains, still occasionally sleety, destroyed the illusion created by brief, blazing, sunny spells that summer had already arrived. A small tribe of bullfinches had appeared from nowhere to devastate the buds amid the white blossoms on my plum trees and, by mid-month, a flock of white-fronted geese who had wintered on the marshy area at the far end of the loch were flying high, heading north-west to Greenland and their breeding areas. Liane, as is usual with wildcats in the mating season, had lost her appetite and was in the woods more often by day. I hoped she was meeting Sylvesturr but could find no evidence of his presence on once-weekly searches. Her 'fast' lasted just under three weeks, then she was back on the window sill at nights, *mauing* to come in and chirring noisily at every touch of my hands as she slept on my lap at the desk.

By now I was busy on eagle treks for I had been given licences not only to photograph at an eyrie but to find and monitor all the eyries over a 185 square mile area. On one ten-mile foot slog up a long green valley, feeling dwarfed by the near 3,000-foot ridges of the mountains on either side, we saw nine large stags filing upwards over the slopes. Four still had their antlers, four had only bleeding stumps where their antlers had been, and one looked comically lop-sided with just one left. When I tried to take tele-photo pictures of them, the shimmering heat haze from the sun annoyingly blurred the images. We found a new fox den, two buzzards perched by an old refurbished nest in an alder tree, saw a tawny owl fly from her nest in a riverside stump, but not a single eagle eyrie did we find.

CHAPTER 12

Moobli was most protective towards little Clarrie and often lay down beside her allowing her to do the same.

One day in early May I was boating up the loch to check some rock faces to the west, Moobli running along the shore for hard exercise, when I saw him stop short. A large ewe then appeared over a small tussocky ridge, glaring at him, and stamped her hoof in anger. Moobli had been well trained to ignore sheep so I knew he hadn't stopped on her account, but it was unusual for a sheep to run a few steps towards him aggressively like that. Now he went up onto the ridges, well away from her, scenting something and apparently staring into the woods. Something odd was happening.

Suddenly I saw the browny-black form of a hill fox leap up the far bank of a deep burn and disappear between the trees. Moobli had got the fox's scent and instinctively started after it but knowing he hadn't a chance to catch a fox with a two-hundred-yard start in woodland, and not wanting him lost for hours, I whistled and called him back as I beached the boat and ran up the ridge. The ewe then ran away but all at once a lamb picked itself up from the early sprouting bracken on the ridge and ran straight into the water, tottering weakly from side to side. It flopped about as it tried to swim out. I dashed for the boat and, flailing away with the oars, caught up with the lamb and lifted it aboard. It was bleeding from the throat where the fox had clearly attacked it, and seemed almost all in, its snowy-white fleece dripping wet, eyes half-closed as it tried to bleat feebly for its mother. I looked up but the ewe had vanished from sight.

Although the wound did not seem to be bleeding too badly, the fox's teeth had torn a deep hole just above its front left shoulder so that its trachea was exposed. It would not recover without treatment so I took it home quickly as Moobli sat panting in the bow after his run. As I squeezed the water from the lamb's fleece and warmed it against my chest, Moobli stared with tongue-lolling smiles as if he thought he would now have another four-legged playmate round the cottage. He seemed proud of himself. Well he might be, I thought, for if he had not been running along the shore I would never have known the lamb was there or that it had been injured, for it had been behind the ridge and I probably would not have been looking in that direction anyway.

Over the years on our long treks Moobli had scented no fewer than five sheep in trouble, caught up in bramble bushes, sometimes on ledges in burn gorges which I would not have seen as I passed by yards above. Sheep trapped like this can starve to death or be drowned after heavy rainfall causes rushing spates in burns. Yes, I thought jokingly, maybe I'll be sending the farmer a bill for Moobli's services!

Back home I mixed warm milk and water in a sheep feeder bottle and was glad to see the lamb could still swallow and though very weak would take the bottle. I got two-thirds of it down her, bathed her wounds in cool water – not hot for that would have stimulated the blood flow – spread anti-biotic cream over them, then put her in the rear workshop with plenty of grasses, hay and kale from my garden for bedding and food. She was so weak, however, she could not stand, just lay with her head on the hay, eyes closed, jerking her legs occasionally. I felt sure she would not survive.

But little Clarrie, as I christened her, put up an amazing fight for life. Next morning, as I left for another long eagle trek, she was on her feet and had eaten much of the grass and kale. I fed her as much warm milk and water as she would take, renewed the food supply and left with Moobli to spend the night out in the hills.

After an early morning hunt for eagles which produced one unused eyrie, I boated home fearing the worst. But Clarrie, now as dry as a fire peat and her curly fleece fluffed out, bleated loudly and ran towards me as I opened the door. My previous toilet problems with Liane were reduced to total insignificance in my memory as I saw the mess now in the room. Clarrie, it seemed, had a unique ability to convert half a pint or so of watered milk into some two pints or more of fluid. And that was only one half of the story. I decided to let her run outside in the fresh air as it was sunny and warm. But because I did not want her to totter off and maybe get lost and die and also because her wounds were weeping, I closed her in the wildcat pens where there was plenty of vegetation for her to eat. A good way to keep the pens cropped too, I thought. She moved about, clipping off the herbiage with both upward and downward jerks of her head, nibbling so fast and continuously it made my jaws ache to watch. In less than twenty-four hours she had cropped all she wanted, mainly selecting bramble and hazel leaves, sorrel and several grasses, but leaving buttercup, hogweed, foxglove and the leaves of little rowans.

My slight worry now was how Liane would react to the presence of a lamb around the cottage. While it is often printed that wildcats will take small lambs I'd never heard of it happening in my area. But I felt Liane was still runty for a wildcat and Clarrie was a hefty little lamb of about seven weeks old, so there seemed little danger apart, perhaps, from the blood scent of her wounds. To my surprise I did not see Liane for two days. The previous night's food in the woodshed disappeared but all the milk had been left.

I went round the woods with Moobli but although we did not find her, the rear haunch of a newly-dead deer calf had been stripped of hair for a few inches and some meat eaten, far more neatly than by any fox. She would not have left the area without good reason. Perhaps she had seen the strange big white animal in her pens and this had triggered the wild streak in her to assert itself at last.

By now little Clarrie was quite tame, bleating at the sight of us, quite unafraid of Moobli, and would come running for a drink when she felt like it. But she was also curiously independent and once fed wanted to wander off on her own, always heading upwards. She was still unsteady on her feet and I did not want to risk her being caught by any big fox again, so every few hours I had to go out and search for her. Luckily her blazing white woolly uniform was easy to spot. I carried her in my arms down to the front pasture again, which she accepted quite stoically, head lolling back with eyes closed, nuzzling into my beard as Liane had done when younger.

On the night of May 13th, after being away three days, Liane returned to the woodshed, hungry and so thirsty that despite the burns and lochside she drank nearly half a pint of milk. I wondered if she had found Sylvesturr in the long wood and with Cleo gone had perhaps made him an offer he could not refuse! But now Clarrie had vanished. Oh well, she would probably be all right for one night. I would go out and fetch her back early in the morning.

I had no need to, however. I was just dressing the next day when a shadow appeared at the window. Moobli, ears pricked forward and an anxious look on his tan face, was blocking out the light by rearing up like a small bear on the outer sill. I went out and he whined, leading me into the kitchen. There, sitting on the spot where her haybed had been in the rear workshop, was Clarrie, calmly chewing the cud! She had remembered the place where she had recovered and now regarded it as a good shady place out of the sunshine to sit for an hour or two. She had obviously ambled past Moobli as he lay on the hall mat, which he did every morning before I got up, and all he had done was to come and fetch me! That day I tried tethering her with a collar – until I found it caught up on a small oak tree and Clarrie lying where she had fallen while trying to pull away. She was breathing stertorously as the collar was too tight on her neck and she'd been too daft to sit back to ease the pressure.

For the next few weeks life with Clarrie involved me forgetting her for a few hours, rushing out, seeing no trace of her, then dashing to the heathery ridges four or five hundred feet above, where she would be quietly grazing, her head stuck deep into the grasses. She would suddenly see me, bleat plaintively *'mair air air'* in greeting, then allow me to catch her and bring her down again. She needed constant watching now, and to have a pinch or two of aureomycin in her milk bottle feeds, because the wounds were festering and needed daily treatment.

I soon found my fears that Liane might attack her to be unfounded. Twice when the front door was closed we found Clarrie escaping the midday heat of the sun by wedging herself into the shade of the gap between the woodshed and cottage walls. And Liane was scared of *her* – so scared in fact she walked up and down *mauing* but would not go into the woodshed while the lamb was so near her entrance hole!

Occasionally, when Clarrie was grazing near the cottage, Liane stalked carefully towards her through the blooming masses of bluebells, making sure not to go nearer than four yards; with her head going up and down like a badger scenting when first emerging from its sett, she peered at the odd, white but apparently amiable creature that had suddenly invaded her life and spent most nights in *her* pens.

Not until May 22nd, however, did it appear Liane had finally accepted the stoic lamb. I was typing at my desk when I noticed that Moobli, who was lying outside, had his ears pricked forward as he watched something by the cottage wall. I sneaked out – to see Clarrie standing by the pens and Liane walk slowly out from the woodshed gap and very gingerly, most tenderly, sniff Clarrie's nose. Then, as if overcome by her own daring, she suddenly danced away with dangling tail back into the shed. Clarrie seemed not one whit alarmed by the wildcat's nearness and calmly toodled off up the north hill, grazing as she went.

By now we had a guest staying at the cottage – Geoffrey Kinns, who lived in London and worked as an artist at the British Museum of Natural History. I had come to respect him enormously. To see snow-haired, bespectacled Geoffrey walking with his almost indiscernible limp down a city street, hear his rather quavering voice and observe his diffident self-effacing manner, one would perhaps take him for a bank clerk, or a civil servant who had spent his whole life in a city office. How wrong one would be. Our friendship had begun three years earlier when on a visit to the Museum I had noticed superb photographs of golden eagles above his desk and he had informed me he didn't think they were very good but that he hoped to take better ones next year! Since then I'd come to realize he was one of the finest wildlife photographers in Britain, totally unsung publicly because of his natural modesty and because he regarded such work as his 'hobby' – so much so that he often virtually gave away his pictures, which have appeared in nearly every worthwhile nature book in this country over the past twenty-five years. The majority of the photos in the authoritative *Handbook of British Mammals* are his. Geoffrey had photographed, often in the wild, almost every mammal, amphibian and reptile that exists in the British Isles. When I say 'almost' I mean he had not for instance yet got round to photographing the rare sub-species of shrew, *Sorex araneus granti*, which lives on the island of Islay!

On holidays to the Highlands he had spent seven seasons photographing golden eagles at the eyrie, had numerous sightings of wildcats in the wild to his credit, and was relentlessly determined, patient and inventive when securing his pictures. Once, to photograph a rare wild pine marten, he had on the spot rigged up an astonishing tripwire device on a branch along which the marten often ran – using a wedding ring as a pulley! All this he had achieved against handicaps which would have put most men off for life. He had served with the 8th Army right through the North African campaign, had survived the push through Europe after the landings in Italy and had been about to engage in the crossing of the Rhine when his left leg was blown off by a mine just below the knee. Despite this, and suffering from high blood pressure and a weak heart, with his doctors constantly telling him to take life more easily, such was Geoffrey's love of nature and his wish to communicate love and true understanding of wildlife to others that he was still at the age of fifty-five going to night classes on natural history subjects three times a week and each year carrying a 45-pound pack up steep mountains in the Highlands in his pursuit of the eagle.

He was only staying with me for a few days before heading up to try and photograph eagles in Skye. I felt honoured by his interest in my wildlife work through the year and by his friendship, and I naturally valued his opinions. He told me I was crazy to regard Liane as a runt. 'Wildcats vary greatly in size, as you found when you examined the pelts at the Museum,' he reminded me. 'She's a superb specimen. Do you mind if I take some cine film of her – it may help wildcats in the public eye?'

An unassailable argument, so for three days we filmed Liane. Once again, now she had an audience apart from plain familiar old me, Liane behaved as if she was trying for an Oscar. She danced about, played with Moobli, sniffed Clarrie's nose again and began catching young slow worms from a family born beneath my lumber pile. Once she came down from the north hill with a dead vole in her jaws and started playing with it near the woodshed. I had never been lucky enough to photograph that before myself, I realized, as I dashed inside for *my* camera. After a little more play she started sunning herself, panting in the heat like a dog.

One afternoon, by tying small pieces of meat on the branches of a large old hazel bush, we filmed Liane climbing over them and the tree's mossy bole. Then she played around on the bank below amid a riot of wood sorrel, violets, primroses, celandines, dark green dock leaves and bluebells. After the fourth time Liane became bored and headed back towards the woodshed round the front of the cottage.

There was a sudden loud scream, the like of which neither Geoff nor I had ever heard before. One of the common gulls from the island colony just off my beach had landed on the bird table after a crust. And Liane, who had just emerged from the herbiage, had leaped after it, missing its tail by an inch. The gull had uttered the screech knowing it had suffered a close shave with death and never came back to steal from the small birds' food again. Unfortunately the incident happened too fast for us to film it.

One morning Geoffrey wanted to film Liane coming down the north hill in long shot, working her way back to the camera en route to the woodshed. I carried her up to the first ridge amid the brambles and bracken warning Geoff he would only get one chance as she was bound to run when set down in the open. When he shouted he was ready I put her down, then ran to be out of shot and threw myself

Liane never played with half-alive prey, though she often did once it was dead. Here she clutches a field vole with both paws before flinging it high into the air for some practice pounces.

earthwards – straight into a bed of brambles. As the hooks pricked my face and stuck in my arms and legs, I waited a while then yelled 'Did you get it?' No answer. Three times I shouted but still no answer. I got up and peeped over the edge. Geoff was by his cine camera staring at Liane nearby.

'I didn't see her in time,' he explained as I came down. 'One second she wasn't there then she bolted down so fast I couldn't get onto her.' I caught Liane and carried her back up the hill, but much the same thing happened the second time.

'What's wrong with your eyesight?' I shouted in sudden exasperation, as I came down again picking the prickles from clothes and skin. Geoff just looked – and I withered. It was a damn fool thing to say to a man who had been Battalion Shot in the Hampshire Regiment! I apologized and we tried again. This time Liane went as a wildcat would normally in open country, sneaking down cautiously from cover to cover, and Geoffrey captured it all.

Next day we drove over to Kingussie to the Highland Wildlife Park where they had some eight wildcats in a large enclosure. While I felt they would not breed in such a situation, it was fascinating to see so many versions of Liane living apparently amicably together. As some passed others that were lying down by rocks and bushes in the sun they made the same *'Brroo'* trills in the throat I had so often heard Liane make, or replied with sharp *'Maoos'* if they did not feel like fraternizing at that moment. When the time came, I would certainly find a mate here for Liane, I thought.

When Geoffrey had left for Skye I dug the composted garden over ready for planting, getting bitten by the first midges, and had just gone back to the cottage when I saw Liane race down past the window. She dug a hole in the bare earth, performed her toilet while watching all round her nervously, raked it all over again, then fled back to the woodshed. So that was where she was going. She clearly liked the soft dusty surface. Again I recalled how the other wildcats, old Sylvesturr included, would bury their droppings on sand, shingle, dusty earth or snow but not on grass or old bracken or harder surfaces.

By mid-June, working at it part time and with the help of two fine young keepers, I had managed to locate and check seventeen golden eagle eyries in my 185-square-mile area, in which a total of four eaglets were being reared. One of these had been found by accident by two local men, Allan MacColl and Stewart Thompson. Although Allan kept sheep he admired eagles and Stewart kindly agreed to become my helper. We had erected a hide of totally natural materials some forty-five feet away from the eyrie on a cliff ledge which, ideally, had the cover of small dwarf trees growing from it. I had to go thirteen miles by boat, drive twelve and walk in six miles on each return visit. Naturally, I had little time for other work during those weeks but I obtained the best pictures and observations so far of these birds – and the eaglet flew successfully in early August. Out of my ten visits, because I wanted to observe eagles at night and dawn, I spent five nights in the hide on the ledge while poor old Moobli had to wait in the Land Rover where the road ended. But he bore it stoically, looking forward to his exercise run along the lochside when we returned home to Clarrie and Liane.

The little lamb slowly gained in strength and on the days I was home often barged in through the front door, wandered about the kitchen knocking my home-brew beer bottles over – the sudden racket made her perform odd skittish leaps into the air – chewed chair legs, nibbled at the white walls and often flumped down beside the dozing Moobli to chew the cud. But while she often went up to the high ridges where adult sheep passed near her, evidently none of them was her mother. She became more independent, however, and by mid-July refused to suck milk, finding all the sustenance she needed from natural grazing.

Coming back from the eagles one evening I went to where I had released Cleo and her other three kits but did not see any of them. I met the keeper of the land, who said the rabbit population in the fields seemed to have decreased and that one night a wildcat had been seen on top of a fence post near the road. From his description, and knowing Cleo sometimes liked to climb onto a pole or stump to have a good look around, it seemed she was still surviving well. It was good news.

Encouraged by this, I went back a few days later to keep a lookout some fifty yards from the disused wildcat den, spending the night in the Land Rover. At dawn I was watching the rocky cairns by the roadside, tired from the hard nights out on the cliff ledge too, when I suddenly saw a plump young wildcat emerge from the herbiage, look nervously about, creep along a few yards then vanish back into the undergrowth. I tried to take a telephoto but found the eyepiece of my camera had steamed up completely from the night cold. Then to my astonishment another young wildcat with a thinner face came out, followed almost exactly in the tracks of the first and also vanished. I felt sure the first one was Eeny and the one behind was the slim wiry tom Miny. I feverishly cleaned the eyepiece, wanting to see if Cleo or Meeny would also show themselves but just then a small van went zooming by as if its driver was late for work and although I waited another half hour, I saw no more of them.

I could, I suppose, have set the box-cage live trap for them but if at least three of them were surviving we had done well enough. Besides, I had plenty on my hands right then with the eagle work and Clarrie and Liane at home.

CHAPTER 13

Liane stretching in the sun as she relaxes after cleaning herself.

Liane was now well into her second year and had become extremely active. Sometimes I saw her at dusk chasing insects and large moths across the ground, leaping up into the air after them as if warming up her skills before going off on the hunt proper. She was firmer, plumper – not yet as large as Cleo but, like her, all muscle – and she was obviously hunting well for she was occasionally leaving my meat in the woodshed and staying away two days at a time. Wildcats are not fully mature until two years old and it seemed she was into a new growth phase for she now weighed over six pounds.

None of this weight increase, I was glad to note, was due to enlarged milk glands and I had to admit that even if Sylvesturr *had* come back around the area months earlier, or she had met him in his own wood in early May, he had certainly not made her pregnant. I say 'glad' because I had by now made arrangements for a rather odd trip – and if Liane had been expecting kittens I would have had to cancel it.

Since a book about my years in the Canadian wilderness had been published in North America, I had received letters from several notable folk, including two distinguished ecologists. One was Dr Victor H. Cahalane, author of the definitive *Mammals Of North America*, *Alive In The Wild* and other fine works, who had been associate director of the New York State Museum and chief biologist of America's National Parks. The other was Sigurd F. Olson, former President of America's National Parks Association and also of The Wilderness Society, who had written eight memorable books about the wilds, including *The Singing Wilderness*, *Runes Of The North* and *Reflections From The North Country*, and who lived in Ely, Minnesota. Both the latter were well into their seventies and had invited me to visit them, as also had Jim Doherty, editor of *International Wildlife* magazine in Milwaukee. All these men lived in the eastern states but I also intended to meet others in Vancouver, on the Pacific coast, as well as revisit my little log cabin and the wild haunts of my Canadian years.

Ten years earlier when I had first driven right across Canada in an old milk truck into which I had built a caravan 'home', I had vowed that if I was still capable in my fiftieth year, I would repeat the journey. What cheaper way could there now be of meeting everyone yet travel freely across the vast continent without the worry or cost of hotels? All I need do was buy another old milk truck, install a couple of shelves, a gas cooker, water supply, buy food and a piece of plastic foam for a bed and drive away. Through a pen pal called Anne Chapman who lived in the backwoods of Wisconsin and always signed her letters 'Woodswoman Annie', I located a good 1970 milk truck at the Golden Guernsey Dairy in Milwaukee whose directors agreed to sell it at a very fair price.

A friend who worked on a neighbouring farm and had already been correctly introduced to Liane agreed to boat down to Wildernesse at least once a week to feed her. He took Clarrie to join five other orphan lambs on the farm itself, I left Liane a large food and milk supply in the shed, and with Moobli staying with friends he knew and liked in Worthing, Sussex, I set off in late August.

It was an extraordinary journey, for the old truck only had a top cruising speed of 46 m.p.h., but after scary brushes with rattlesnakes in the broiling deserts of Montana, battling snow blizzards in the heights of the Rockies, and a dusk encounter with a wild grizzly while temporarily inconvenienced in some bushes in Glacier National Park in Canada, I did manage to meet everyone on my schedule. By the time the old truck had traversed the American and Canadian Rockies and shuddered back to Chicago and a new owner, I had put 5,400 extra miles on her clock, mostly in eight and a half days of gruelling sixteen-hour drives. Although I had left Liane when hunting was at its easiest and I knew my friend would not let me down, my thoughts often returned to her as I wondered how she was faring.

When I arrived back at the loch with Moobli in the Land Rover, exhausted and looking forward to quiet days at home, my friend told me Liane had stopped eating the food he had set out for her during the third week, and in fact he had never once seen her.

With the boat fully laden we biffed up the loch against the waves from a strong easterly breeze. I let Moobli out for a short exercise run, and as I unloaded belongings he panted up to join me in the walk to the cottage. All was total peace at Wildernesse. A family of woodcocks shot up from a clearing in the bracken and winged their way into the west wood. The bramble bushes were loaded down with blackberries but only a few yellow spikes of bog asphodel, blue scabious and pink wild roses were now replacing the flowers of summer. I hastened to the woodshed and to look into the den in the pens.

Of Liane there was no sign.

Next day we searched both the woods but apart from fox droppings full of blackberries near an old deer carcass below the west wood, Moobli could find no scents at all. I wondered if perhaps some boating holidaymaker had taken Liane but as my friend, who had held and fondled her before my departure, had not even seen her, this seemed unlikely. She was surely too wild and sensible, once she knew we had gone, to allow a total stranger to pick her up.

Three days later we boated up the loch again to post urgent mail and fetch the rest of my gear from the truck. Visiting my friend again, we were delayed and had to come home in the moonlight, but as I would be indoors catching up with my writing for the next few days I let Moobli off about two miles from home for a hard run.

As I watched him loping along like a great wolf in the silver light over the rough, boggy, rocky and tussock-filled ground where a man could hardly walk, I thought I saw a dull-coloured animal like a large hare suddenly dash down the hill into his wake, chase him for a few steps, then dart back up into the bracken again. I blinked, shook my head, looked again but nothing now moved. How odd, I thought. It was probably just a trick of the light or due to the few whiskies I had consumed with my pal.

Waking late at about 9 the next morning, I wandered drowsily round to the woodshed feeling sure that Liane, who had been staying away a day or two at a time before we had left, had now gone for good.

But suddenly I heard a loud *mauing*! She was back in her hay box bed and hearing my step and voice, and seeing my looming shadow at the door, leaped out with a loud *'Brroo'* chirr, sharpened her claws briefly on a slanting log, then started walking about me with loud caterwauling cries. She was thinner than before, her eyes standing out from her head in disbelief at the sight of us.

The grey-brown figure that had dashed out one and a half miles away the previous night had clearly been Liane. Only by Moobli running along the shore, by his scent and perhaps the exact memory of the engine note of my big boat, had she realized we were back and had followed us home. She reached up and buried her claws painfully in my thigh, then leaped into my arms. I fed her all the raw and tinned meats and milk she would take after which she fell into a deep sleep in the hay box. It was then I noticed some small dog footprints in muddy open spots a few yards from the shed. Perhaps someone with dogs had visited the area and that had driven her away, for it seemed strange she should leave the area in which she had been born. But perhaps she had gone to *look* for us? It was odd, too, that she had stopped eating the good food my friend had left out, for that meat was now teeming with sexton beetles.

For the next six weeks Liane seemed permanently hungry, *mauing* loudly for food and eating as much as I did each evening, and half as much as Moobli. Where she packed it all I had no idea but each morning her bowls were empty. She quickly replaced her lost weight, then stacked on more as she laid on her winter fat until by late November she tipped the scales at $7\frac{1}{4}$ pounds.

Despite her raucous welcome, however, she now seemed far more wary and wild when we caught her out in the open. But when she came to the window sill at dusk on October 5th and caught me in the act of pouring some milk into a mug of tea, she saw the familiar carton and *maued* loudly, although she had just drunk her normal bowlful in the shed. Encouraging her, I let her in and was just pouring out some milk when she turned and knocked the mug of tea flying. The hot liquid touched her feet just before I picked her up to get her out of the mess and – as I'd often done when she was a kitten – threw her across onto the bed. But she spat loudly in mid-air, landed, then crouched looking angry and cruel, her ears back, eyes narrowed, snarled at Moobli and continued to growl. It appeared her long time alone in the wild state had caused this sudden reversion. I knew better than to react with resentment or even try to mollify her, so I told Moobli to stay where he was and ignored her. A few seconds later she jumped onto the window sill and I quietly let her out.

I just hoped this odd setback would not last.

This was now the stag rutting season – when the big master stags come down from the high hills where they have spent the summer together and seek their 'harems' of hinds, and Scottish estates with unenclosed lands cull by shooting the numbers suggested to them by the Red Deer Commission. It had been a fine long summer and the stags seemed to be late this year, for we had only heard the usual rutting roars twice. One sunny day Moobli and I set out up into the hills to try for photos and to see if the main numbers had come down yet from the high tops.

After a mile we saw a small group of eleven hinds on the slopes to the north-east but without stags. Then another group to the east –

also with no stags among them. Turning north-west we came to the 1,500-foot peak overlooking a huge green glen below Guardian Mountain and found a larger herd of twenty-one below us, but they had no stags with them either. And all the rest of the glen was empty. Feeling we were still too early, we came off the peak and headed south-west over an undulating plateau filled with deep tussocks, their flowing bent grass crowns turning orange before the final pale sere yellow of winter. Here and there we crossed small drier tables of higher ground where close-cropped heather vied with patches of short bracken now in its autumnal yellow stage. My intention was to check a badger sett on the slopes above the lochside where two cubs had been reared that year, but we never reached it.

Suddenly I noticed the large white heart-shaped rump and grey-brown back of a solitary deer grazing, head away from us, in a fold in the ground. It seemed too large for a hind, then I saw the tops of the antlers moving like dead branches tossed by a wind as the head made cropping movements. Indicating to Moobli to keep back, for he was well trained to hand signals now, I crouch-stalked to the south-east until I reached a small gully – I didn't want to go so deep I couldn't see its legs – then stalked back westwards into the wind again. Readying the camera, the sun now shining perfectly onto the stag from the south sou'west, knowing it would soon look up and probably stare before running, I got to within forty yards then sank down to one knee. It looked up all right – a magnificent beast, though its ten-point antlers were in-curved and not large – then kept staring as I took shot after shot.

To my surprise, far from running, it even moved nearer, showing its cloven hooves perfectly! I kept clicking away as it moved half to the right, half to the left, looking back disdainfully over its shoulder – for it still couldn't make out what the odd immobile shapes were. Finally, wanting a broadside shot, I started talking to it, telling it thanks very much but it could now go away if it liked! And finally for good measure, a couple of skyline shots as it ambled away. It was not my idea of an *ideal* stalk, for you've only done your job properly if you get your pictures without the beasts ever knowing you were there, but it was good for just a few hours' work.

With the sun still blazing down, I had a flannel bath in the kitchen, cut my hair with the aid of two mirrors on a camp bed outside, then

By late November in her second year Liane weighed $7\frac{1}{4}$ pounds, still runty for a wildcat. Here she cleans herself in a rare sunny spell.

lay down to sunbathe. A few minutes later I felt a tickling sensation. Liane had padded round from the woodshed and had installed herself in the warm space between my thighs. She squirmed and rolled, twisting her head upside down and her movements tickled so much I could hardly bear it. But feeling this should make up for her reversion in the study a few evenings earlier I let her please herself, bit my lip and stuck it. When I tried to pick her up to bring her onto my chest, however, she *maued* as if annoyed that I had interrupted *her* idea of the way friendship could be expressed out in the open, and when I let her go she ran back to the woodshed.

It was clear now that the separation during my longest-ever absence from the Highlands had induced the return of some of her wildcat instincts. Like a youth called up in wartime, she had realized it was a harsh and lonely world when one had to rely on one's own inner resources, and that independence made one strong. She had, in short, grown up. The period had done her good but from now on our friendship would be on her terms as much as mine.

CHAPTER 14

Liane, now weighing over eight pounds, gives further proof that she likes fresh carrion, clawing at the rear haunch of a red deer hind that I had found dead in the woods. She growled and clawed at me when I removed the leg.

The winter that followed was the coldest, snowiest but also the driest I had experienced in eight years of Highland wilderness living. But the blizzards and constant cold seemed not to bother Liane at all. Her coat grew thicker and darker and she often spent her nights out in the thick hay of the den in the pens, only returning to the woodshed in gale-filled or rainy periods.

The cold did not worry Moobli or myself either, despite the nearest gas and electricity outlets being over six miles away. With a 44-gallon diesel oil drum, a discarded asbestos chimney pipe and a day of hard banging with a cold chisel and hammer, I made a cottage heater for the commercially produced equivalent of which, one of those fashionable new wood burning stoves, one would have had to pay upwards of £200. Laying the barrel on its side over rocks, sealing the chimney at one end with a paste made of powdered asbestos and water, and cutting a rectangular fuel door at the other end, hooked on tightly with a piece of zinc sheet, I had a superb heater. It kept the study warm for hours with just two smouldering logs. And of course wood heats you three times anyway – once when you cut it, once when you split it and again when you burn it!

In November Liane still seemed to be eating so much food I wondered if perhaps Sylvesturr was joining her at night to help clean up the bowls. One evening Moobli gave his loud baying 'fox' series of barks and started off towards the west wood. But wondering if he had sensed or scented the big wildcat – he never took to the fierce old tom as he had to the females – I called him back.

Over the next two days we found fresh wildcat droppings in the east wood, far too large to be Liane's and certainly not those of a fox. The fresh carcass of a small deer calf had been chewed neatly from the rear and the flap of skin had been carefully raked back over the exposed area, almost certainly a big wildcat's work. Moobli could only follow the scent for a short distance before it petered out heading west along the loch shore. I was sure now wildcats can obscure their foot scent by choice of route, treading in the water itself, on hard twigs and along fallen logs and rockfaces. If it *was* Sylvesturr I did not want to scare him away again by setting the live trap, especially if he was also visiting Liane to take some of her food. On the second night I noticed Liane's bowl was only half empty yet her stomach was distended – something had to be taking what she left behind.

There was one safe way to find out – keep watch. I closed both doors of the woodshed, then placed a hazel wand with a wooden square at one end as a 'gate' near her entrance hole. Then I retired to the dark of the workshop. My plan, if I saw the big tom go through the hole, was to sneak out and block it up, thus trapping him in the shed, so I could check his condition and maybe photograph him. But although I spent several hours this way on three nights not a shadow of the old boy did I see. On two other nights I just went out at 1 am and closed the gap anyway but all I saw was Liane gazing at me sleepily, wondering what on earth the sudden new visits in the middle of the night were for! Then there were no more big wildcat scats in the woods. It seemed Sylvesturr had eluded me yet again.

In early December Liane gave further proof wildcats like fresh carrion. After a period of heavy south-westerly gales hail and sleet storms when the loch level had risen five feet in two days, requiring both boats to be manhandled up the shore out of reach of the waves, Moobli and I found a freshly dead huge hind deer in the long wood to the west. Her forequarters had been chewed by a fox, the stomach removed, but as the big rear haunches were fresh I skinned them and took them home as extra meat for the animals. Then I hung them from a rope across the roof beams in the shed. When I went out later Liane was standing with her rear feet on the hay box, her front claws embedded in one of the haunches, and was fiercely chewing away at the red meat.

The blood taste and smell made her wilder and she growled when I lifted the leg away, kept clinging to it, her body taut as if made of elastic. When I moved her hay box so she could no longer reach she actually leaped onto the haunches when she felt peckish and chewed large holes in them. And when I put her onto my shoulders for a normal stroking session she used me as a high perch, snaffling a few more bites as I was holding her, and even growling slightly. Such rich food seemed to make her wilder and I was sure if I'd fed it to her all the time it would have been harder to tame her as much as I had. As I held her that way I could actually *feel* the power of her jaws as they bit into the meat and I was extremely glad she was friendly. She entered a new growth phase too, and by mid-January weighed just over 8 pounds.

For the rest of that winter, Liane ran totally free but remained semi-dependent on us for food. Days went by and we barely saw her, then suddenly she was *mauing* away on the window sill again, seeking to come in for reassuring contact and play sessions which were now conducted with gloved hands. Then having assured herself all was well with us and the status quo had not changed, out she went again to pursue her own life.

In early March, with the missel thrushes singing me awake at dawn from the highest larches in both woods, the great tits beginning their rusty bicycle pump *'Tee-chirr'* notes, and the chaffinches experimenting with the first bars of their trippety courtship songs, Liane came into her second-year oestrus. She was clearly feeling the urge to mate far more strongly than the previous spring, for she constantly rubbed herself over twigs and grass tufts, rolled on the ground with her head completely upside down, oblivious of its wet surface, performed little somersaults, and made loud churring growls as if with frustration. She even walked about growling on half-bent legs, her tail high in the air, backside quivering sideways, and tried to 'stand' for Moobli. And the dopey old dog stood there with a big lop-sided affectionate grin, like a great horse over its foal, enjoying her sudden matiness without the slightest idea of what was causing it. She also went into the wildcat's usual springtime 'fast'.

But I had anticipated this, though I had expected it to occur a few days later, and I knew just where I would borrow a mate for her. Were not the eight wildcats at the Kingussie Highland Wildlife Park somewhat overcrowded, and were they not well used to human beings? Surely the Park would let me have a surplus tom for a few weeks, maybe sell me one for good if he and Liane got along?

On my next supply trip, however, when I telephoned the director Eddie Orbell, I received a shock. 'We *would* have let you have one,' he told me. 'But the blizzards and heavy snowfalls this winter broke down the enclosure and all the wildcats escaped! We set traps of course but were only able to get three back – two females and one tom. Obviously, we can't let you have him!'

So as I returned home, chiding myself for not having checked up earlier, I realized it was probably the end of any chance of a mate for Liane that spring season, unless the fact that she often peered intently towards the west wood at dusk now meant that old Sylvesturr *was* still paying visits, and with Cleo gone might just mate with her. I felt, however, this was unlikely but I was not unduly worried. Liane was still a young wildcat and it would do her no harm to wait until she was more mature, in her third year. After all, it makes us appreciate the good things in life more if we have to wait a little for them!

By March 24th, still fasting and eating only half her normal rations, it seemed Liane's oestrus period was over. She came in through the window for the first time in several days but now there were no loud chirring *'Brroos'* when she was touched, just a few twitches of her tail, and she was calmer and more placid. Within a few days she was playing like a kitten again during sunny periods, often in the open pens. She darted to my hands through the mesh, clasped them gently between her paws, claws always sheathed now, gave a gentle love bite or two, then hared off again to stalk rustling grasses, crouching down so low, her pupils round and black, she looked like a thrown wet rag, her body fitted so closely into the contours of the ground. Sometimes she threw herself at one of the five-foot posts, hauled herself to near the top then stayed clinging there with all four sets of claws to be tickled on the neck.

Often, when I threw heavy sticks forty yards in opposite directions to give Moobli hard exercise, she would watch them flying through the air, her head going all over the place as if torn between the desire to see and the desire to flee. Occasionally she made token runs after Moobli as he pounded past and up the north hill, but as he thundered back with the stick she scooted into the safety of the shed gap. She seemed even more interested when I threw a round rubber ball for Moobli, so once when he'd had enough and was sitting down panting heavily in the front pasture, I rolled it towards her. To my surprise she darted after it, gave it a few round-arm cuffs, clutched and rolled over it looking more like a large monkey than a cat. It was the first time she had played with a ball but she soon lost interest in it.

By April 1st she looked slimmer, more long-legged, and I found she had lost a pound of her winter weight since January. Two days later she began her spring moult, losing hair with almost bare patches appearing on her fore-shoulders, under her limbs and her front flanks. And her new growing coat was a much lighter colour, the guard hairs buff and yellow.

I thought by now her oestrus period was certainly over but on April 5th, when I had to go out to meet a keeper who lived twenty miles away in connection with eagle work, I saw her again trying to 'stand' for Moobli. Well, it was just a slight reoccurrence of the mating urge, I felt, called Moobli down to the boat and we set off. It was late at night when I returned to the anchored boat, a heavy south wind was blowing and as it was almost pitch-dark I decided to sleep in the truck until daylight.

When we returned to Wildernesse Liane was nowhere to be found. Some daffodils near the woodshed had been snapped off their stalks and her food and milk tins were still half full. Had someone come and taken her? I searched the area, found no footprints apart from my own, then went round both east and west woods with Moobli. There, in a muddy patch near the west wood, I found a deep but obscured track which appeared to be four-toed. But the mud was so soft and fallen in I could not tell whether it had been made by a large wildcat, fox or a dog. Moobli, whose own tracks were twice the size, picked up a scent but it petered out once we were in the dry needle-covered floor of the wood. Then we found a large new twisted scat filled with deer hair near the rhododendron bush by the path, and another way up on the north hill. Again, it was hard to establish whether they had been left by a wildcat or a fox, for they seemed too curled at the ends to be those of a dog, and too large to have been Liane's.

When she did not appear on the second and third day I became really worried. A couple of days and a night had been the longest she had ever stayed away before, apart from the time we returned from my North American trip. With Moobli I then hiked along the loch-side and up along all the ridges and rockfaces beyond the west wood, continuously imitating the wildcat *mauing* calls. Suddenly I thought I heard a faint answering *'mau'* coming against the wind from a cairn of large boulders to the north-west. I looked at Moobli. His ears were pricked forward and he was looking in that direction. He had heard it too.

We climbed up and Moobli found a scent, zig-zagged between the great tangle of rocks, where again it appeared to peter out. But there on a mossy boulder was a similar scat to the ones we had found near the cottage, though this one was not fresh. I was now almost sure they had been left by a wildcat, possibly Sylvesturr, but when I continued the search, *mauing* loudly, there was no answer. If Liane *was* with Sylvesturr in one of the several possible dens in the large area of the rockfall, perhaps he had a way of telling her to keep quiet! After another hour of searching without result we came home, and I put fresh meat and milk in her bowls.

Next morning, a Sunday, I woke up after hail had been hitting the tin roof much of the night to find a blanket of snow covering the ground. I felt sad and gloomy, sure Liane had gone for good. If she stayed out in this weather, having not been fed for four days, she would probably become too weak to return even if she wanted to. I walked round to the pens – no tracks in the snow and no sign of her in the den. Even the sight of the first primrose and two celandines in the snowless lee area behind the cottage failed to alleviate my depression. The shed also looked dead and empty. I looked over the half door – some of the meat and all the milk had gone. I checked the door to see if Moobli had pushed his way in, but I remembered he had been in the kitchen all night.

Then I heard a long drawn out *'Mau'*. I looked up. Liane was sitting like a silver ghost on the largest log of the woodpile in the dark at the back of the shed! I could hardly believe my eyes, it was like seeing an old friend come back from the dead. I *maued* in reply but she looked scared and dropped to the floor. As I opened the door she started to flee for the exit hole as if I were a total stranger. But I stayed still, *mauing* gently, and finally she came slowly towards me. She looked odd, frightened, and was much thinner although she had left half the meat. There was a tired look about her too. It took a full five minutes to warm her wet body up with my hands and get her purring again. I felt a great joy and relief, realizing now just how fond I had become of her, and it was wonderful to touch her long soft fur again and feel her larynx vibrating against my own throat with her loud purrs.

I felt then that if she had been with Sylvesturr, or maybe even another tom that had migrated to the area, he might just come back again to visit her. If so I wanted to catch him and establish the fact beyond doubt. I put Liane into the pens, closed the gate and set the baited old box-cage trap on the far side of the rhododendron bush out of sight of the cottage.

Next day, the trap untouched, I let Liane out again for a few hours but at dusk she was sitting by the pens and peering intently towards the west wood. Afraid she would go off again or enter the trap herself at night, I shut her in. I repeated this for several more nights, allowing her free periods each day when I and Moobli, who would soon have

In long and varied play sessions with Liane, Moobli has the patience of a bear with its cub.

let me know if she strayed too far, could keep an eye on her. But when the trap remained untouched, no new droppings appeared, and she ceased to gaze at the woods, I let her run free again. I was sure her oestrus period was now completely over. I realized too that if the droppings we had found were those of a fox and not a wildcat, Liane could just have gone away to *search* for a mate and not necessarily have been with Sylvesturr or any other tom. She could have been acting on mere instinct.

These theories seemed to be correct for during the next five weeks Liane showed no inclination to wander. But on May 15th I was surprised when she came into oestrus again, for the *third* time. That morning she started growling, moving sensually over grass tufts with her tail high, and making loud *'Brroos'* as she once more tried to 'stand' for Moobli whenever he came near. Two days later after I had stayed out overnight working on eagles, Moobli and I came home to find she had disappeared again. At first I was not too worried as she had not been very hungry when she had returned after her second absence, and the weather was now fine. But on the fourth day I set out with Moobli early in the morning, when scent is strongest in the dew, and tried to track her through the woods, pasture and the slopes and ridges above in a half-mile circle. This time he found no traces whatever and looked uninterested as if he knew she was miles away and had gone by choice. Nevertheless, I kept setting her food out each night in the woodshed.

Liane was nearly two years old, strong and healthy after a life of good feeding, and I was sure now that she had been desperate to mate and had gone deliberately to look for a tom. If this was true it was most interesting that female wildcats would go in search of males in this way, as was the discovery that her oestrus period had erupted three times, the second after a gap of twelve days, and the third occasion after a lapse of some thirty-five days. If this happened with unmated females in the wild, and they *mated* during such a third oestrus, the average of 66 days gestation for the wildcat means they would give birth around July 21st. While it would not be ideal timing, it would still give a *good* mother time enough to rear her kits and teach them to hunt for themselves before the onset of the real Highland winter in December.

I was now not only busy with eagles but also studying some foxes and had to spend the night of my birthday, May 25th, several miles away. When Moobli and I returned next day, the first drizzle for ten days now falling, I had just about given up hope of Liane returning but went to check the woodshed from old habit.

She was back again, this time in her haybox. She greeted me with a loud chirr of welcome and, despite having polished off all the food and milk I had left in her bowls, leaped out and started calling raucously for more. Again she was thin and had sheep ticks on her brow and behind her left ear, but within ten days she was back to her old weight. As the weeks passed by there were no more recurrences of her oestrus but she was not pregnant either, so it seemed clear she had not mated with Sylvesturr, if he was still alive, nor any other wild tom.

I suppose it is possible that one day Liane will leave us, especially if she finds a mate in the wilds. I have given this much thought but have decided against trying to live-trap a wild male for her, not only because it would mean keeping it in unnatural and miserable imprisonment but also because there would be no guarantee that under such conditions the two would mate anyway. If Liane hears and wants to obey the call of the wild, then that is her right. Until that day, if it ever comes, this is her home.

Despite these little dramas, our friendship with Liane has continued to flourish. Moobli spends long periods lying down near the cottage watching her, and she has devised a hilarious new game for them to play. She walks out towards him slowly until right beneath him, turns on her back and cuffs his indulgent face, turns back on her front, stays still for a while, then without warning goes off with a mighty scuffing of earth, so fast he can't catch up to her. Then she comes capering round the other side of the cottage, her body high off the ground, tail like a fat dangling streamer, and darts into the gap by the woodshed to greet his tardy arrival with another playful cuff. But if he becomes too boisterous the claws come out, she stamps and spits and he soon backs off.

I now firmly believe wildcats are 'psychic', for if ever I have to go away for a few days, such as when working with eagles, on overnight treks or on New Year camp-outs, she comes to the window sill the night before. Then she insists on coming in to be stroked, looking up into my face and thrusting her chin up into my beard with little lip-smacking noises, making her chirring notes of affection, just to establish we are still good pals.

Occasionally, especially if she has been hunting alone in the woods all day, she will still come in and sleep overnight on the bed covers beside me, curling up in the warm bend between my legs and body. With adulthood, her early toilet problems no longer seem to occur and she wakes me before dawn with loud purrs and *maus* to be let out again. Perhaps in this and some other ways I may have expected too much of her too early.

When we keep a wild animal we must always try to understand their instincts, anticipate their fears, forgive their lack of understanding, for just as our concept of God is to us, so we must act towards the animal kingdom beneath us. And perhaps this applies not only to animals but to birds, trees, plants and all of nature and what little is left of the inspiring natural world. We are custodians or we fail in our duty.

It is a glorious day and the sun is beaming down upon the loch, the shimmering reflected light dancing upon the fans of new leaves in the wild woods, warming us with the promise of summer's heat to come. Liane is in the open pens again by choice and is basking in the bracken of her 'sun bed' on the den's roof. As I walk towards her and she hears me coming, she lies on her back to look at me upside down, stretches full length until her head is over the edge, pushes both her black-padded front paws towards me as if in greeting, then suddenly falls off! She lands on her feet, not at all put out by the drop, as if she was clowning about on purpose, stretches up as high as she can to give the wooden pillar a good working over with her claws, then gives a loud pleased *'Brroo'* as I tickle the long buff fur of her belly.

In this moment my mind's eye sees again the runty little kitten being licked clean by her mother just after her birth, being kicked away by the larger stronger kits as she tries in vain to reach her first teat. I recall the difficult, hilarious and rewarding moments of the long training weeks in the cottage, and I feel an odd mixture of pride and gratitude that although Liane is free to go where she will, with miles of uninhabited wilderness behind her, she has so far chosen to remain with us. It seems a true if unusual friendship, and it is one that will never be taken for granted.

Liane basking in abandoned fashion in her sun bed on the roof of her den.

A fully mature, happy wildcat called Liane.